The Geography of Uncertainty

This book outlines the characteristics and implications of a potential geography of uncertainty. In doing so, it analyses this concept in reference to both the origins of uncertainty in Early Modern Age and the current geopolitical situation.

The book adopts an interdisciplinary approach to uncertainty, drawing on global perspectives and literature to define its meanings and characteristics. In order to develop a thorough and precise understanding of the geography of uncertainty, a broad perspective is adopted, which includes other forms of knowledge in which the concept of uncertainty is firmly established. As such the book creates temporal links, that may occasionally be far off from one another, to present a geographical perspective of uncertainty. It provides an interpretation of the phenomenon of globalization in a new way, relating it to the first European openness to global spaces, the Early Modern Age, and identifying the transition from the medieval world to the Modern Age as the first manifestation of uncertainty in geography. Uncertainty is more prevalent than ever in today's geopolitical, economic, financial and social reality, as well as the ongoing emergencies and crises.

The book adopts an interdisciplinary approach rooted in the geography of Early Modernity by referring to geopolitical scenarios, literature and philosophy to target the historical roots and the prevailing configuration of the geography of uncertainty. It will appeal to scholars and students of human and political geography, politics, philosophy, international relations, economics and history.

Alessandro Ricci is Assistant Professor of Political Geography at the University of Bergamo, Italy. His research interests concern Early Modern globalization and political geography; historical and political cartography; the connections between art, maps and power; and the Netherlands during the Golden Age. In 2018, he received the "Carmelo Colamonico" award from Accademia Nazionale dei Lincei for "scientific writings in geography". In 2014, he obtained a European Label PhD degree *cum laude* in "Culture and Territory" at the University of Rome "Tor Vergata", where he has been a researcher and attended courses in political cartography and geopolitics. He carried out research activities at the University of Trento and Amsterdam. He is one of the managers of the Think tank "Geopolitica.info".

The Geography of Uncertainty
A Conceptual Model of Early Modern Globalization and the Current Crisis

Alessandro Ricci

LONDON AND NEW YORK

First published 2024
by Routledge
4 Park Square, Milton Park, Abingdon, Oxon OX14 4RN

and by Routledge
605 Third Avenue, New York, NY 10158

Routledge is an imprint of the Taylor & Francis Group, an informa business

© 2024 Alessandro Ricci

The right of Alessandro Ricci to be identified as author of this work has been asserted in accordance with sections 77 and 78 of the Copyright, Designs and Patents Act 1988.

All rights reserved. No part of this book may be reprinted or reproduced or utilised in any form or by any electronic, mechanical, or other means, now known or hereafter invented, including photocopying and recording, or in any information storage or retrieval system, without permission in writing from the publishers.

Trademark notice: Product or corporate names may be trademarks or registered trademarks, and are used only for identification and explanation without intent to infringe.

British Library Cataloguing-in-Publication Data
A catalogue record for this book is available from the British Library

ISBN: 978-1-032-49513-2 (hbk)
ISBN: 978-1-032-49516-3 (pbk)
ISBN: 978-1-003-39420-4 (ebk)

DOI: 10.4324/9781003394204

Typeset in Times New Roman
by Apex CoVantage, LLC

Contents

	Preface	*vi*
	Foreword	*viii*
	Introduction	*xii*
	Acknowledgements	*xvii*
1	For a definition of uncertainty	1
2	The society of uncertainty	20
3	Early Modern European political geography and uncertainty	41
4	The "mad flight" and the geography of uncertainty	69
5	Cartographic secularization	83
6	The tragedy of cartography in the Modern Age	102
7	Uncertainty as a paradigm of Modern Times	121
	Bibliography	*130*
	Index	*144*

Preface

Discourse on Journey

Only a few geographers remember that the great ambition of human and *vidalian* geographical knowledge, the one which was taught until recent years in our universities, was to produce a geography of feelings: fear, hate and love. One century and a half ago, the enunciation of such a programme forced the last heirs of Vidal de la Blache to give up such an impossible task, but they could indeed conceive it abstractly, convinced as they were of its necessity: the geography of the most impalpable and immaterial, evanescent, existing things, which was different, let's say, from the geography of investigation which regarded the shape of the roofs and houses or the altimetric limits of crops that they were educated to.

The achievement has now been attempted by a young geographer, who, boldly and inattentive of every border, crosses the most different fields (from political history to logics, from epistemology to philosophy to anthropology and to the history of literature) chasing one of the most elusive and fleeting, but at the same time crucial, conceptual entities: the opposite of what we call certainty. And his itinerary lies in such still unmapped lands (19th-century explorers' uncharted land) and cannot rely on any already-beaten path, on any sign left from some other explorer, on any "invitation" placed at a bifurcation by those who preceded him, as happens when you walk in the mountains or in the desert. Alexandre Koyré wrote that nothing in historiography has had a more nefarious influx than the notion of "precursor", because to indicate someone as such, inevitably involves the inability to understand him. But such a statement, which has a retrospective meaning, also applies to its opposite, prospectively: woe to not report promptly he who ventures in an unknown territory without a guide, because sooner or later, we will be inevitably following in his footsteps. And, therefore, the maximum attention must be dedicated, from the beginning, to journeys like the one that follows. After all, the term "method" literally means precisely this: something whose existence is only possible "after the journey", that is after that the pioneers have opened up the road. Pioneer indeed: the foot soldier who, in previous centuries, preceded the bulk of the army and opened up the way with an axe, a hoe and explosives.

No wonder, then, that along the road that awaits the reader, some passages are not always firm and are still waiting to be fixed, some bridges would have everything to gain in being a little firmer and so on, as with such really perspicuous, in this case, expression is said. The author takes his risks (as he must) so that those who come next can proceed more safely and directly. And this is the only way of building, not only the famous method (that is always a multiple and collective elaboration), but before that, a tradition. We have an urgent need for this and that, and in some way now we can no longer draw back from the work, now that the track has been opened. Evidently, Alessandro Ricci, as all those who practise music, knows something that Vidal's pupils did not know: that if you can think of a note, you also know how to play it. This is why he finally succeeds in doing what French classical human geography could only think of doing.

Franco Farinelli

Foreword

In autumn of 2010, after a thesis defence in Modern History about Capitalism in the Netherlands during the Golden Age, I started to collaborate with prof. Franco Salvatori at the University of Rome "Tor Vergata". Due to my enthusiasm, he asked me if I wanted to give my contribution to the didactic activities, starting from the topics dealt with during the elaboration of my master's degree thesis. He was convinced that the best way to start didactics was to throw one in at the deep end and see if he would sink or swim. I substituted him for a couple of lessons, due to his busy agenda as pro-rector and president of the Italian Geographical Society, in a course with a limited number of students, so the impact would have been adequate for a young scholar. The title of that course was very challenging and fascinating for me: "Geography of uncertainty and information". In the days before my first lessons, I focused my attention on the first part of the title, trying to collect some useful insights for the students. I understood that that topic would commit me for the next few years.

Some of the more experienced colleagues tried to warn me that I would have had some difficulties in elaborating that research topic, considered too general if not too ambitious. When I put these doubts to prof. Salvatori, he clearly told me that a research topic, even for a young researcher, *must* be ambitious, because a scholar *must* pose great questions and *must* have great perspectives.

During the period when I stated to concentrate on this topic, global geopolitical order was facing harsh blows: after the 9/11 attacks, the new millennium had been characterized since the first years by radical changes in the way of interpreting world order, the war, the economic waves, the international assets and, at the same time, by a new use of force, a public violence, in a continuous state of emergency that was seen by many authors as the new political paradigm of globalization.[1]

After the terrorist attacks in the United States, American military involvements firstly in Afghanistan, in 2001, and then in Iraq, in 2003, contributed to changing the perspective on the only global superpower and on its capacity of immediate and effective intervention in each and every part the world, which had been one of the main certainties of the "globalized order" created in the post-Cold War period. The 2007–2008 financial crisis, with its following waves on a global scale, shocked the "second certainty" of globalization, the capitalistic pillar of liberal democracies and the linked global order.

In some way, those two great events corresponded to a new phase of the geographic international system, in which uncertainty emerged as a key factor in global politics and economy. The Arab Spring was also a rising political phenomenon in the Middle East and North Africa (MENA) region, which then involved European migration policies, shocking the established and precarious – we have to say – order that had been established in the area for decades. That movement, that was both external and internal to those national contexts, exacerbated the ethnical, tribal and identity struggles in some of the involved countries. It destabilized political assets and produced crises, the consequences of which strongly impacted European stability and the ability to face the migration crisis, on years to come.

The Syrian and Libyan civil wars, the emergence of the Islamic State (IS) in 2014 – which was established as a strong regional and religious power during the internal struggle in Iraq and Syria – also due to the American intervention, represented another challenge for global order that was constructed after the end of the bipolar system.[2] IS was an international actor which contributed to destabilizing the European and a strong part of the liberal world: with its terrorist attacks and its attempt to replicate a Caliphate, it attempted to establish a new (or renewed) *global* order, based on religious and political patterns at the same time.[3] This international vision, well-expressed in IS's propaganda and in its symbolic perspective, was the most evident symptom of a particular threat to the Westphalian order and to the political geography that it had founded. Indeed, IS violently proposed to prevail over the international divisions established on territorial assets and boundaries. The Caliphate caused a crisis in the main global geopolitical divisions and ways of intending international relations, contributing also to the creation of chaos and destabilization in the attacked countries, in particular in the regional contexts in which its provinces were declared, creating new boundaries and uncertainty in political maps.[4]

In substance, the historical moment in which I started to ponder these topics was well-suited to a complex, long period and global discourse, which was what I wanted to propose.

The economic crisis seemed to throw the world into a period of chaos and instability: new geopolitical struggles were emerging in the MENA region, the Balkan tensions of the 1990s had not disappeared, the global war on terror had not ended, the Eastern European countries were characterized by deep transformations and the Arab Spring was opening up a new phase of destabilization for the MENA region.

The reflections on those geopolitical changes led me to make comparisons and parallels with the Early Modern period. I noticed that the instability and uncertainty the entire world was living in after the Cold War assets (that were unstable, but certainly more *certain*) were for some reasons similar to the Early Modern ones.[5] In both cases, there had been a transition from a certain system to another, which needed new certainties and new political and existential pillars. In both cases, global vision and globalization emerged as key topics to read the world changes.

Therefore, when I started to ask myself what the origins and the deepest roots of that uncertainty were, my answer was immediate: if globalization was emerging as the most characterizing word of the post-Cold War period, and if that period was clearly appearing as uncertain and founded on terror and emergence, the early steps of globalization had to be the origins of modern uncertainty. In other words, if I wanted to understand the contemporary geopolitical uncertainty, I had to look for the answer in Early Modern globalization processes and in the geographical perspectives of that period.

Late Renaissance, which included the end of medieval certainties and a new phase of cartographic representation, perfectly embodied the rising uncertainty. Indeed, that period was strongly characterized by a paradoxical uncertainty: man acquired new certainties in the scientific field, in geographical knowledge regarding new continents and lands, but all of these new, certain discoveries corresponded to the rise and affirmation of uncertainty in public and private lives, which is what I will try to explain in the next pages, starting from the geographical field. Indeed, the globalization process started with the first uncertain travels of Cristopher Columbus, Amerigo Vespucci and Ferdinand Magellan, which were markedly characterized by uncertainty of destinations and lands to be discovered. They all symbolically opened up the Modern Age. It metaphorically corresponded to the emergence of the Nation States and to a global vision of international relations and to a precarious and uncertain balance of power, as well-explained by 16th-century authors such as Niccolò Machiavelli and Giovanni Botero.

Baroque times were also interpreted as the affirmation of tragedy, which was the theatrical interpretation of uncertainty of the modern human condition and of cartographic representation. This apparently reached new scientific and certain methods of representing the world, but it was at the same time, the most evident consequence and element of modern uncertainty. I will try to demonstrate this by also referring to the most famous characters of modern tragedy, who apparently characterize uncertainty in the modern human condition.

The conceptual framework of Modernity was a continuous affirmation of a paradoxical uncertainty that originated with geographical discoveries and an opening up of the world, which is the intrinsic main factor of globalization.

Lorenzo de' Medici, not by chance, targeted the most relevant topic of his times, just two years before Columbus's travels, when in 1490, he wrote the famous lyrics "Quant'è bella giovinezza,/che si fugge tuttavia!/chi vuol esser lieto, sia:/di doman non c'è certezza" ("There is no certainty in tomorrow"): he perfectly caught the end of certainties of late medieval times and rising Renaissance.

***

When the book was first published in the Italian version, in 2017,[6] the international scenario was strongly characterized by the uncertain, chaotic, revolutionary and critical events mentioned earlier. *Crisis* seemed to be affirmed as an intrinsic characteristic of the new world order, in which on the one hand, financial dynamics greatly contributed to global uncertainty and to the inability to prevent future uncertainty and plan national and international policies (Casti, 1991). On the other hand, the global political theatre emerged with so many critical scenarios that it seemed impossible to establish an order and points of reference able to give collective and personal certainties. Better to say, it appeared clear that the new world order was directly connected to *crisis* and *uncertainty*, and that the globalization order was paradoxically – intimately – disordered.

The book received some attention at that time and – also due to the particular attention paid to the topic – I started to perceive *uncertainty* as an emerging and relevant subject to interpret geopolitical relations and assets, with books and public events created around it.[7]

The international events of the last five years, the COVID-19 global crisis,[8] public debates on apparent scientific certainties, the Russo-Ukranian war which broke out dramatically in February 2022, the uncertainty of media coverage and information and the energy crisis deriving from financial flows and from geopolitical crises convinced me that a renewed and English version of the book was necessary, in order to give the academic international community of geographers, amongst others – in continuous search of new certainties and answers – my contribution to the debate.

I know perfectly well that the topic is challenging and very ample. Therefore, I apologize in advance if I will not give the readers exhaustive, *certain*, complete answers, but this is the paradoxical objective of this book.

In a time of very apparent certainties, my book aims to pose questions and create conceptual links that I hope the reader will comprehend, in a continuous and implicit recall to contemporary times and geopolitical scenarios. I hope this book will contribute to understanding modern uncertainty and to be freed from it.

From the Italian version, the last chapter has been erased and some other minor parts have been inserted, in order to render the book more coherent and based on new literature insights. All the translations from non-English books have been carried out by myself: I hope to have grasped the full meaning of those writings.

Rome/Bergamo, December 2022

Introduction

In the 1970s, a renowned Harvard economics professor decided to quit teaching at the prestigious university in order to embark on a different adventure: a television programme for the BBC. This professor was John Kenneth Galbraith, who partnered with four men of the network staff to write and direct a show, meant to face the major issues of the world economy, addressed in such a way as to allow the audience to understand present and past dynamics, through the history of economic thought and of economy overall – without neglecting the influences that great economists' thoughts exerted on the world's destiny, the great commerce and matters of international politics.

Regarding the choice of the title – according to the author – it was almost instantaneous: *The Age of Uncertainty*. The decision was due to the immediacy of the concept: "it sounded well: it did not confine thought; and it suggested the basic theme: we would contrast the great certainties in economic thought in the last century with the great uncertainty with which problems are faced in our time" (Galbraith, 1977, p. 7). The attempt was therefore to confront the main 19th-century pillars of thoughts – which rooted their certainties in capitalism, socialism, imperialism, and colonialism – with those of the 20th century, a time in which "such certainty" had almost completely been extinguished (Ibid.), even if the topic of uncertainty in economics was firstly faced by Frank Knight (1921).

If this was Galbraith's idea some 40 years ago, more recently, Giovanni Arrighi and Beverly J. Silver (1999; 2001) have argued that the international political situation after the collapse of the communist regimes was characterized by uncertainty and disorder.[9] In their economic and political analysis, the authors started from the ideas expressed in *The Age of Extremes*, where Eric Hobsbawm stated that "the basic units of politics themselves, the territorial, sovereign and independent "nation-states", including the oldest and stablest, found themselves pulled apart by the forces of supranational or transnational economy, and by the infranational forces of secessionist regions and ethnic groups. Some of these – such is the irony of history – demanded the outdated and unreal status of miniature sovereign "nation-states" for themselves. "The future of politics was obscure, but its crisis at the end of the Short Twentieth Century was patent" (Hobsbawm, 1995, p. 11).

In the political situation that followed the Cold War, Arrighi and Silver referred to Hobsbawm's words, considering that the end of real communism "destroyed the international system that had stabilized international relations for some forty years. It also revealed the precariousness of the domestic political systems that had essentially rested on that stability", with "an enormous zone of political uncertainty, instability, chaos and civil wars" (Ibid., p. 10).

This position is clear indeed: according to the reasons provided by Galbraith and later by Arrighi and Silver, among others, contemporary age is characterized by uncertainty, both economic (Lunghini, 2012) and sociopolitical. In fact, as affirmed by many scholars, we live in a society dominated by uncertainty (Bauman, 1995; Appadurai, 1996; Beck, 1992; Benasayag & Schmit, 2003; Chan, 2009; De Cecco, 2007; Veca, 1997), by a fluidity that represents the most common characteristics of economic and financial dynamics (Clark, 2005) and of society itself (Bauman, 2000).

The topic of uncertainty has been addressed by multiple disciplines – from economy to sociology, from history to philosophy, from math to psychology[10] – especially during the past few years, when the critical financial dynamics contributed to exacerbating the dimension of personal and collective instability and also with the rise of some geopolitical crises, which extended beyond national borders and seemed to spread internationally (Kissinger, 2014).

Geography appears not to have been particularly involved in such conceptual considerations, due attention has not been paid to the topic of uncertainty neither from a geographical economic perspective nor from a geographical political one. A few researchers so far have studied the concept of chaos, or of uncertainty in a broad sense, mostly from perspectives[11] of critical geography and geopolitics. Thrift (1989; 1992; 2007; 2005) proposed an analysis of international relations of recent years from the point of view of the chaos and disorder that seems to dominate them, while other authors, like Kenichi Ohmae (1990; 1995a; 1995b) and Ken Jowitt (1992), Michael J. Shapiro (1991; 1994), Claudio Minca and Luiza Bialasiewicz (2004) have focused on the consequent idea of indefiniteness of the current borders, also studying them from the viewpoint of the uncertainty originated with the end of the Cold War and the consequent rise of East European nationalisms. This subject was also addressed by Robert D. Kaplan (1994), who stated that "the borders . . . have become largely meaningless" (Ibid., p. 16) and by Henry Wai-chung Yeung (1998), who contested the idea of a *borderless* world (see Ohmae, 1990; Ó Tuathail, 1999), arguing that the territorial and national dimension linked to the borders continues to have an irrefutable relevance even in economic relations. A somewhat similar idea has been addressed by John A. Agnew, a supporter of the idea of the "territorial trap" (Agnew, 1994), and by Kaplan a few years later (2012) with "the revenge of geography", both affirming the leading role of the geographical element in consideration of international relations. Others have analysed the topic of chaos in the modern geopolitical situation (Ramonet, 1998; Rampini, 2015), referring in particular to the dynamics of deterritorialization and of progressive – and alleged – fluidity and the "erosion" of international borders, especially in the past 20 years.[12]

In tourism and regarding the *Non-places* of Marc Augé, Minca has applied a similar metaphor to the touristic flows of recent years, speaking of "ephemeral spaces" (Minca, 1996). According to the author, who has focused on the study of post-modern (Hassan, 1987; Wellmer, 1985) in geography (Dear, 1988), "the touristic space, in other words, tells us about the new, uncertain relationships established between society, individuals and the context in which they are protagonists" (Minca, 1996, p. 187).

Other geographers have focused their analysis on the migratory dynamics and their effects on Italian cities, highlighting the chaotic changes and the consequent dimension of uncertainty (Cattedra and Memoli, 2013), a viewpoint that seems similar to a wider economic literature related to the concept of uncertainty arising from migratory flows (Todaro, 1969; Harris & Todaro, 1970; Langley, 1974; Hart, 1975; Rogerson, 1982; Gordon & Vickerman, 1982; Maier, 1985; Berninghaus & Seifort-Vogt, 1991).

The purpose of this work is to try to outline characters and meanings of a possible, potential "geography of uncertainty", a notion that does not only refer to the current geopolitical condition but mostly to what produced it, its origins. In order to try to finally come to such a definition, I will start from a definition of uncertainty drawing on international literature, adopting an interdisciplinary approach. It will be imperative to broaden our vision to other forms of knowledge and seek temporal connections, at times distant from each other, in order to formulate a complete and clear idea of the geography of uncertainty.

***

An analysis of the concept, from a geographical point of view, is hence proposed, relating it to the first European openness to global spaces, and I will try to identify the transition from the medieval world to the modern one as the first manifestation of uncertainty in geography.[13]

Regarding the topic of crisis of the Modern Age, vast literature can be accessed,[14] as shown in the following chapters, so I will try to focus on specific emblematic aspects of uncertainty in the geographical scenario, trying then to let the possible connections with today's international reality emerge.

Where to retrieve the reasons, the origins of geography of uncertainty?

First and foremost, a connection with European openness to global spaces, to the geographical extension that has occurred since the Age of Discovery; the uncertainty perceived in current geopolitical, economic and financial, social and international relations scenarios appears to be the result of a long-term process initiated in Modernity and thanks to geographical knowledge and certain geographical-political dynamics that was triggered in the European continent in that given historical moment.

Modernity represented, as a matter of fact, a total upheaval of political, conceptual and philosophical categories of those times: the whole structure of the previous era – that could be defined as ideologically certain – entered a state of crisis (Le Roy Ladurie, 1980).

That phase, characterized by systemic changes concerning the international scenario, the relations among States, the European way of thinking and world political geography, is identified as the pivotal, starting moment of uncertainty in the geographical *discourse* and in its representation, starting from the loss of those certainties guaranteed by the Middle Ages, at least from a religious point of view.

In order to reach a definition as organic as possible of modern uncertainty and to apply it to the geographical area, it will be necessary to start from other disciplines – both humanistic and not – that have already solidly addressed the subject and provided vivid images of Modernity and inter-State relations, not only in a political and sociological perspective but also regarding economic relations and the influxes that they had in the social sphere, precisely in terms of uncertainty.

It will be particularly useful, not only for the reader but also for he who is writing, to restrict the reflections on the subject within a framework of general perspective, at least as regards other disciplines, and to later verify if some of the other authors' theories could be replicated, adapted or "overlapped" onto the geographical discipline. And eventually, it will be necessary to analyse in which ways this could be possible. I will therefore tap into a rather heterogeneous bibliography. Being this work of a theoretical nature, I will mostly use secondary sources pertaining to different disciplinary and temporal backgrounds, which could contribute in providing argumentative and reflective support in a geographical perspective.

The methodological choice of sources was based on some main considerations: the authority of the cited authors, their influence on international debate and the capability they had, in several respects, to address the topic that here I want to lead back to the geographical epistemological field.

After explaining how researchers today have interpreted modern uncertainty from multiple perspectives, I will try to verify whether it is possible to talk about uncertainty in a geographical context – both effectively and conceptually – analysing probable implications in Early Modernity: in the loss of medieval certainties and in the uncertainty of political order within Europe, the redefinition of its borders, in the light of the key concepts of *chaos, general crisis* and *revolution*.

I will try to trace some lines from the origins of uncertainty in the Modern Age, predominantly starting from the great voyages of exploration and from literary expedients of the accomplishment of the "mad flight" (referred to Colombo's journey), to use Dante's metaphor, intended as the beginning of the "epistemological tempest" (Turco, 2010, p. 277) – and not only – proper to Modernity.

A focus will be given to cartographic reflections of the advent of modern uncertainty, trying to define a "cartography of uncertainty"[15] (or, in other words, what destiny belongs to cartography as the expression of this world, which is characterized by uncertainty) and the implications of geographical uncertainty in the representation of the world.

Referring to the thesis of those who dealt with the subject of Modern Times as a new irruption of the element of tragedy in history, I will consider some emblematic characters, symbols of the literature of those times, in order to interpret them in a cartographic key, emphasizing the uncertain and contradictory elements of cartographic representation at the beginning of Modernity.

In the meantime, I will try to understand the evolution that geography of uncertainty faced up to the present times, as it is manifested today in international relations. Indeed, it seems that geography of uncertainty could be a fitting definition for the actual phases of globalization and for the early modern ones,[16] focusing mostly on the uncertain dynamics on borders, on political geography and financial fluxes (Cox, 1997).

At a first sight, it could seem very challenging to link the concept of *uncertainty* to the geographical field, but the first objective of this book is to try to understand if this formula can be a sort of synonym of early globalization and then which are its origins, to finally verify if and in which way uncertainty has been lived geographically, at which scales and with which consequences on society and on international politics.

Acknowledgements

I want to thank my mentor, Prof. Franco Salvatori, who motivated me to give my contribution and to deepen my interest in research, even though it was ambitious, global and diachronic. Thanks to my friends and colleagues, Simone Bozzato and Pierluigi Magistri, and to the entire research group in Rome. Thanks to Edoardo Boria, Marco Maggioli, Matteo Marconi and Filippo Menga for the continuous and stimulating dialogues.

Thanks to Franco Farinelli for his useful suggestions and for his dazzling "Introduction" that really honoured me.

Thanks to my new and inspiring Bergamo colleagues for the useful daily confrontations and common human and professional pathways shared, in particular to Federica Burini, Renato Ferlinghetti, Alessandra Ghisalberti, Manuel Anselmi, Massimiliano Bartolucci, Agostino Brugnera, Paolo Cesaretti, Riccardo Fanciullacci and Domenico Perrotta.

Many thanks to Angela Landi for her very precious collaboration in the text review, and to Carlotta Bilardi for her contribution to the translation and for her constant support.

Also, thanks to my students, from the first course to the present ones: without the insights attained during the lessons, without continuous confrontation with them, I would never have published this book.

Thanks to my friends and colleagues of "Geopolitica.info", in particular to Gabriele Natalizia. Thanks to my family and my loved ones, for their support and warming presence.

Thanks to Routledge and in particular to Faye Leerink for their collaboration and openness.

In conclusion, thanks to all those who have had to put up with my continuous and reluctant uncertainties.

Notes

1 See Agamben, 2005; Benigno & Scuccimarra, 2007; Colombo, 2022.
2 See Ricci, 2018a.
3 See Ricci, 2020.
4 Ricci, 2015b.

5 A similar perspective of parallel between Modern Times and contemporary reality was also proposed, among the others, by Hardt and Negri, 2000; Benigno & Scuccimarra, 2007 and Sloterdijk, 2013.
6 See Ricci, 2017. See also Id., 2010; 2013a; 2020b; Ricci & Salvatori, 2019.
7 See Fusco et al., 2017; Valori, 2017; ISPI, 2017; Senanayake, 2021; Foreign Affairs, 2022; Colombo, 2022. Conferences on: *Emergence/Emergency* (Rome, 2021); *Mapping Uncertainty. Early Modern Global Cartography* (Rome, 2022); *La città dell'incertezza* (Rome, 2022); *Managing Uncertainty in a World in Transition* (Rome, 2022).
8 See Ricci, 2020a; 2022.
9 See Arrighi & Silver, 1999; 2001; Barnes, 1999; Hannay, 2008; Horsman & Marshall, 1994; Jowitt, 1992; Kennedy, 1993; Rosenau, 1990; Thrift, 1989; 1992.
10 In the 1990s, a workshop on psychology and work considered in the perspective of uncertainty was held at the University of Padua. Prof. Vittorio Rubini wrote in his premise that it could have been possible to refer to uncertainty as a place into certainty, as something precious, to be protected even when we have certainty (in Bellotto & Foppolo, 1997).
11 See, among the others, Agnew, 1994; 2009; Cattedra & Memoli, 2013; Minca & Bialasiewicz, 2004; Ohmae, 1990; 1995a; 1995b; Shapiro, 1991; 1994.
12 See Amin et al., 1982; Bauman, 2000; Berman, 1988; Khanna, 2011; Ó Tuathail, 1999; Thrift, 1989; 1992; 2005.
13 Relevant thinkers of global history will be taken into account, also in a diachronic way and sometimes considering only specific relevant aspects of certain historical periods or great authors in respect of the subject matter that is the one of uncertainty, or more specifically, of uncertainty in geography and its causes.
14 See Aston, 1965; Hobsbawm, 1954a; 1954b; Kamen, 1971; Kossmann, 1960; Parker & Smith, 1978; Trevor-Roper, 1959; 1967.
15 With this term, I am not referring to the interesting and valid studies regarding uncertainty in actual systems of cartography and GIS techniques, which is a very different topic (see Atkinson & Foody, 2002; Hunsaker, 2001; Mowrer & Congalton, 2000), but to a more theoretic and paradoxical way of representing the world that was typical of the Modern Age.
16 See Braudel, 1979; Hardt & Negri, 2000.

1 For a definition of uncertainty

Geography and uncertainty: oxymorons?

The question concerning the relationship between geography and uncertainty, so seemingly distant from each other, is: how can we draw a first parallelism, a draft of rapprochement between the two?

This is fundamental in defining the categories that identify and approach the concept of uncertainty. It implies a fundamental paradox which must be investigated. Geography is indeed considered an "exact" science. It is practically impossible to conceive it as indefinite or uncertain, and, most of all, it is voted to the *orderliness* of things. So, what is this oxymoronic juxtaposition between the two terms and where does it come from?

It is appropriate to start from the meaning of the word *geography*, from its etymological root which derives from *γεω*, *γῆ* (Earth) and *γράφία*, from the verb *γράφειν* (to write) which indicates a writing of the Earth, a description of the world.

What kind of writing is it? Is geography a purely descriptive discipline or does it more widely set goals that can go beyond the mere writing of the world? Also, if it is to be considered a description, it must surely lie within a cultural framework; a framework defined as semiotic from which we must draw in order to create the description, the "writing" of the world. Does this mean that we will have definite or indefinite representations according to the cultural context in which they are formed and developed, in which they are elaborated?

Ptolemy had defined geography as an imitation of the design of the whole part of the Earth: a description that cannot ignore figurative and representative apparatus of which cartography is the most vivid expression. According to Ptolemy, the descriptive-textual and the visual-representative systems represent the two pillars of geography; knowledge that must take the universe into account together with the laws that "lead" the global world and which are necessary to have order as opposed to chaos.

Elements of writing, based on different culture cannons and according to different perspectives, also regarding power, are needed in order to describe the world. They are never neutral, but activate certain behaviours and presuppose political actions, so that the same description of the world and the corresponding cartographic representation are never neutral tools, but tools that are always intended to put the things of the world in order (Branch, 2014; Harley, 2001; Wood, 1992).

DOI: 10.4324/9781003394204-1

That is why Emanuela Casti has come to define geography as an "ordered representation of Earth" (Casti, 1998, p. 18), hence, a discipline able to give an order through its writing and its representative capability. This order is pursued through ever-different timings and instruments, because each social group follows its own modes and rhythms, generating a mobile path where certainties and doubts are mixed (Ibid.).

In geography, *certainties* and *doubts*, *vagueness* and *uncertainties* inevitably interconnect, so that geographical science itself becomes "mobile", since it refers to social contexts that change over time, changing references and ways of describing and representing the world (Agnew, 2002). What does not change is the prospect of *ordering* the Earth, of controlling geographical space, of carrying out projects and establishing social relationships between man and space (Raffestin, 1980). Hence, geography's main objective, achieved through the use of its instruments, is to put order in natural dynamics. This represents an intrinsic necessity of mankind, in some way *ancestral*. It is also one of the reasons that lead Franco Farinelli to define geography as humanity's primordial knowledge, which precedes philosophy (Farinelli, 2003).

Geography is order and it aims at an order to create the world's representative models, because it follows criteria that presupposes an order and on which it is based. It is a discipline that involves the development of some sort of "inventory of the world", of classification (Nicolet, 1988): not only for descriptive-textual practice but also for a visual-representative one. In fact, in order to describe in words the world or a part of it, or to graphically reproduce it on paper, the geographer has to make choices following a hierarchy of elements. This operation is intrinsic to doing geography, and it shows the ordering properties of this knowledge and its sidereal distance from any idea of uncertainty.

It coincides with the same procedure that a writer, who must properly adapt words in order to give narrative efficiency, and a painter who must choose colours, techniques and objects to produce his work in the best possible way, must use. Both these figures, in their literary and figurative expressions, more or less voluntarily, put things in order to render and depict. They do this according to what is inherent to their sensitive, passionate, critical or social spheres. Still, in both cases, there is a choice to be made.

This also happens – and perhaps to a greater extent – to the geographer who has to choose, according the vast complexity of the world, the elements that he wants to include on paper or in other ways. He does this on the basis of the cultural system of which he is part and on the objectives he sets himself. Therefore, the choice involves determining an *order*.

Under these qualitative geography assumptions, a contradiction seems to emerge in a disruptive and apparently inextricable way: how can geography, which aims to be an ordered description of the Earth and a useful representation which puts nature in order, be approached to disorder, misalignment and *uncertainty*?[1]

There is an inevitable contradiction. But it may only be illusory, at least if the context in which geography is produced is considered. It is necessary, in fact, to understand under a *critical* point of view (see Dalby, 1991; Ó Tuathail, 1996) in which perspective the geographer works, in which world he lives, how he relates to it and according to which prerogatives. If the geographer, or the cartographer, lives in a chaotic global situation, he will represent the world on the basis of that condition. In other words, it can be said that the ways of doing geography, of knowing and interpreting the world, of understanding it and of representing it must be adapted to the time and space in which they are processed. Hence, doing geography corresponds to different historical circumstance and social, geographical and, more extensively, to cultural contexts. Consequently, this latter point derives not only from man's cognitive capabilities at any given moment but also from the historic and the cultural stratifications that have sedimented in that specific place through time, through information gathering and through the most important instrument of knowledge and representation: cartography.

Geography and cartography would then have a supplementary role. They could be seen as replacement models of a lost order or one to be established anew.

On the contrary, if the system were based on "certainties" (ideological, dogmatic, religious and so on), one might assume that the related representative layout was intended to faithfully reproduce the "certainties" in images: it would draw from a semiotic system and a "certain" cultural complex, and therefore, would be an image of them. It should not aim at re-establishing order, because it is the representation of an order itself. That is why in medieval cartography, *mappaemundi* effectiveness and the representative certainties were of secondary importance: the certainties that had to be established were, above all, metaphysical, in a universal and all-encompassing sense.

If the medieval system was devoted to ensuring religious certainties, what happened in European Modernity, when the whole medieval truths apparatus stumbled and fell together with the world of steady dogmatic beliefs?

In the first model, a specific assertive use of solely sacred sources was used: time and space coincided with a single vision that tended to confirm existential certainties. In the second model, established in the revolution of the Modern Age, in a progressive "cartographic secularization" process that from the Age of Discovery brought about some changes in the cultural and social systems.[2] The Holy Scriptures no longer guided the eye, the knowledge and the hand of the cartographer. It was substituted by travel stories, odeporic literature in an affirmation of realism that occurred at the beginning of the 16th-century.

We will move on these assumptions, which are apparently contradictory, and try to define uncertainty.

What does not define uncertainty: indecision, insecurity and vagueness

Trying to define what "uncertain" is appears to be understandably an arduous task, in a certain way. However, in order to do it as thoroughly as possible, and to delineate what is meant by the "geography of uncertainty", it will be necessary to draw from other learnings, different to the one to which I belong, and apply different definitions in order to guarantee a solid foundation and a conceptual framework that is both strong and flexible.

What are the boundaries of uncertainty, if any? Within which limits is it possible to give complete meaning? In other words, what *defines* uncertainty, *gives* it delimitation and subsequently a definition?

According to the encyclopaedic definition, uncertainty is that state of "lack of certainty, a state of having limited knowledge where it is impossible to exactly describe existing state or future outcome, more than one possible outcome".[3] It indicates a temporally undefined condition (past, present and future are involved) of "doubt" regarding something which originates from the impossibility to recognize a univocally accepted truth. This leads to a future consideration in vague, suspended, indefinite terms: *uncertain*.

As evidenced by this brief but effective definition, uncertainty is something that regards men in a non-delimited time due to the loss of truth and universality of "something" that is protracted in an undefined way. This "something" can be replaced by every area of expression of the human condition, geography included, meant as knowledge of the world, as its projection and representation.

Following this path, we can better understand what we define as geography of uncertainty, seen in terms of a loss of absolute truths and the establishment of a system of partial truths, to some extent represented by today's international political "order" – at least in the geopolitical field.

It is an uncertain and undefined order that contributes to creating vagueness regarding the future of international relations. There are many different visions of the word's destiny, even in the short term. They have been well-described in three terms which, from a conceptual point of view, can be approached to uncertainty. They can sometimes coincide and other times have, to a greater or lesser extent, slight differences: *indecision, insecurity* and *vagueness*.

These three notions are tied to the idea of uncertainty, but some differences should be considered, as they will help us reach a more thorough definition.

Uncertainty differs from *indecision*, although there is often concomitance and juxtaposition between the two terms. Indecision, in actual fact, arises after uncertainty, which is the consequence of an impossibility or inability to decide, which takes place in a situation where certainties and landmarks are lost.

Indecision corresponds, more simply, to a *lack of decision making*. It occurs when stability is lost or missing. He who does not take action, who does not move because he can't decide is one who comes from a condition of instability and fears the consequences of his actions. Uncertainty is thus the primary condition, the background where indecision arises and develops. In a state of non-decision, where all useful reference points necessary in making a decision are missing, instability and uncertainty are generated and they form a vicious circle that can only be interrupted by making a *decision* (Schmitt, 2007).

This inextricable concatenation of uncertainty, indecision and again, uncertainty, can be seen both in the field of human existence and in political life, in ideological or in geopolitical contexts, which are of particular interest here. Over the past decades, the question has been the pivot of numerous researches and treaties of all sorts, from politics to economics, from psychology to maths, as we shall see later. Never, or rarely and in a marginal way, in geography and geopolitics.

It is important to highlight, beyond the precise scope of epistemological applications, that the bond between the concept of uncertainty and that of indecision is stringent and bijective: one gives meaning to the other, and vice versa. In other words, both concepts travel simultaneously, although only rarely they coincide.

Indecision, and this is crucial to the end points of this study, is, of course, a matter that concerns the psychological apparatus but, because of that, also concerns relations between States and the geopolitical field. The question of indecision, in politics as well as in literature, finds its completeness in Modernity, where its most representative references, which will be discussed later, are to be found. Indecision is in fact, the symptom of a *crisis*, of a personal or collective critical moment, that necessitates a decision, a stance that avoids the escalation of the crisis. Indecision is hence the attitude of those who face a critical condition: that is a fortiori true if we carefully analyse the idea of crisis, which inevitably contemplates a decision.

Crisis defines a moment in which making a decision is imperative, from its Greek etymological root, "to make a decision": κρίνειν, κρίνω, to separate, to put asunder, to pick out, to select, to choose, but also to judge, to esteem, to prefer, to be of the opinion and, therefore, to make a decision.[4]

Insecurity differs from uncertainty in several essential aspects (Wust, 1985): it corresponds to the lack of security and, according to this elementary definition, the similarity to uncertainty is clear. However, there is a wider connotation which forces us to consider informal but substantial differences in order to move forward. A lack of security always refers to relationships with something else. If this interpretation is to be adopted, insecurity regards social relations and the feeling of vulnerability in relation to danger or uncertain situations. Therefore, to some extent, insecurity – both personal and collective – derives from a condition of instability and uncertainty, but it always refers to someone or something else in relation to the individual or political self.

Michel Foucault (1995; 2009) well describes the process that binds the concept of security and population even in a spatial sense (see Crampton & Elden, 2007): according to the French philosopher, in fact, it is the State that must guarantee securities to the citizens within a spatial system. Security tries to structure an environment, according to a series of events or possible elements, that must be regulated in a polyvalent and convertible framework. The scholar then identifies a crucial point that will recur several times in this work: the need for the modern sovereign to provide citizens with means of security in relation to a territory, and so to order space on the basis of mechanisms which provide certainties, above all spatial which are, therefore, functional.

Security – and in some way certainty – is intended as a necessity for modern political organisms to organize the space of their actions in relation to a population. In other words, the modern sovereign is for Foucault some sort of "architect of the disciplined space" (Foucault, 2009, p. 29). Not by chance then, the French scholar refers to those structuring and political categories, born in the 16th-century, which started the modern political conceptions, even connected to the security and the authority that the sovereign exercises over a territory, identifying two key moments: the collapse of the feudal system and the manifestation of the European crisis due to the affirmation of the Reformation and the Counter-Reformation. He does that starting from a book that he defines "abominable text" (Ibid., p. 89): *The Prince* by Machiavelli (2005), a work that characterizes the modern political conception, as it emerged at the beginning of the 16th-century and that was based on two assumptions inextricably bound to new certainties: territory and population, objects or "targets", according to Foucault, of the sovereign's power.

The modern political system is perceived and characterized by Foucault in terms of a constant search for security, in the relation to territory and population, which is the existential basis of modern Nation States. This quest for security represents one of the modern "governmentality" pillars, as he defines it, in the aftermath of the crisis of the feudal system and which we will call the system of "medieval certainties".

So, if uncertainty is always tied to a relationship between subjects and/or objects and vice versa, uncertainty is something that anticipates this state of things as it is in an earlier ontological condition. The quest for security in the typically modern government *forma mentis* – and here a topic that will be discussed further is introduced – is the symptom of a loss of certainty: it is the result of a lack of stable footholds. We try to provide for this lack through the creation of new certainties, of a new system based on Foucauldian securities.

Therefore, to define uncertainty and to implicitly associate it to the idea of indecision or insecurity is a matter that concerns the geographical field – as we are beginning to notice – which have traditionally been defined in the past as the "relationship between man and environment". It is also, above all, a question which concerns human action in geographical and political terms: the administration of territory and its structures (within given borders) and, on a wider scale, the features regarding relationships between leading entities involved in international politics: the States.

The crisis of the current global order, dictated by several different reasons, is evident, at least as far as the past three decades are concerned, in the geographical and political global apparatus: political, cultural and economic processes of deterritorialization, the close contact between civilizations, the erosion of authority resources and of sovereignty within given borders, and the continual increase of international, transnational or regional institutions and organizations that adulterate the geography of given borders are just some of the most evident symptoms of global uncertainty which afflict the post-Cold War world (Veca, 1997, p. 226; Colombo, 2022). One might otherwise say that the actual and future configuration of the world is seen by man and scholars, in vague terms, indeed uncertain ones.

The third similar conceptualization of uncertainty, even if with different nuances, can be found in *vagueness*.[5] This term has lately been considered mainly in philosophical fields. The first, main similarity of vagueness with uncertainty, as expressed by some scholars, can be found in the impossibility of establishing the terms, the exact boundaries of its own definition (Moruzzi, 2012). It is, in this case, a contradictory and oxymoronic feature: these two concepts, as such, have no *borders, definitions, terms* and they lead us to further terms of comparison.

The second aspect that unites the two concepts is, indeed, that both *vagueness* and *uncertainty*, besides not having possible boundaries, produce existential, conceptual and geographical paradoxes: paradoxes which are the result of disputed truths, of a condition of indeterminacy applicable to the state of a previous system's safety *crisis*.

In both concepts, the paradoxical definition is given by a lack of delimitation.

Wherever existential uncertainty is present, as it will be seen in the analysis of certain authors of other areas of human knowledge, it is possible, if not probable, that this exists in a geographical and geopolitical context too. The *lack of clear borders* is indeed what defines both uncertainty and vagueness, and it *creates paradoxes*.

Therefore, the characteristics of vagueness can be found in uncertainty. Both are paradoxical assets and both produce paradoxes, in the contemporaneity of three further conditions: lack of uniqueness, stability and absoluteness (Ibid., p. 92).

If the product of uncertainty is the creation of paradoxes, a state of uncertainty can politically correspond to two conceivable alternatives: produce a Schmittian decisionism or stay in the sphere of indecision. In both cases, the attempt of establishing safeties, of restoring the modern States' typical certainties, as described by Foucault, is configured. Uncertainty – which is negation of certainties – can therefore be followed by indecision or decisionism, the truth of new certainties or the ultimate lack of them in the approach of insecurity. The inescapable reference is, anyway, always the uncertain state or the instability related to time and space.

In the juxtaposition between *vagueness* and *uncertainty*, there is a feature, which belongs to those contexts in which *chaos* prevails, where the crisis affects a whole system, becoming a *general crisis*, due to jolting movements which are often behind *revolutions*. In such a historical phase, the restoration of new equilibriums and the dismemberment of old ones are certain to take place. This movement, in turn, produces a realignment of political burdens and a redefinition of borders (Colombo, 2014). This is what was witnessed with the conflictual resurgence of nationalities in the Balkans after the fall of the Berlin Wall (Kaplan, 2012), it is what to witness in the Middle East with the clash between different and opposed political models and, from the war in Ukraine as well as from the pandemic crisis of COVID-19, it is what is emerging as a result of the economic crisis in recent years in terms of territorial management and international politics.

Furthermore, the *creation of paradoxes* could be the underlying feature of the observations stated so far, especially in the first paradox to be analysed: the relationship between geography and uncertainty. Another key feature which underlines the important parallelism between *vagueness* and *uncertainty* is that whatever is found in the lack of absoluteness is vague. This is what emerges from indecision, but it also produces insecurity and, undoubtedly, defines uncertainty. With the loss of absolute truths, of given certainties, we enter the domain of uncertainty, characterized by a state of *indecision, insecurity* and *vagueness*.

An uncertainty about which it will be necessary to ask essential questions: where does it have its primordial roots and where and when did it originate?

These first conceptual distinctions, together with the three notions that have just been considered, work as solid support in order to reach a first, even if summary, definition of what the geography of uncertainty is and what it is not.

What defines uncertainty: chaos, crisis and revolution

That moment of deep crisis which developed at the beginning of the 1990s, that politically, economically and geopolitically changed the assets of the world and previously established order, has been compared by Arrighi and Silver – in agreement with Immanuel Wallerstein (1995; 2011) and James N. Rosenau (1990) – to the first Modern Age revolution and to the renewed configuration of the European world that emerged after 1648.[6]

In fact, the parallelism does not appear to be so distant, as in both cases, there has been a shift from consolidated systems of certainties – firstly those of the medieval era and then the political and ideological ones pertaining to the bipolar configuration of the Cold War – to the systems that had lost their reference points and had to be rebuilt on new perspectives.

The core matter of this treaty is strictly tied to such a vision, attempting to foresee a possible connection and interpreting it in the light of the idea of the "geography of uncertainty" of these two periods. In other words, one of the questions at the base of this work is: can present uncertainty, in which, from several points of view, we live, be ascribed to a long-term process, which has its roots in the beginning of Modernity in Europe, in the geographical and political transition from the Middle Ages to Modern Times, that deeply altered not only men's *forma mentis* but also the European political structure?

A change such as the one that occurred in Europe during the 16th and the 17th centuries actually had a disruptive impact on the social, religious, geographical and economic contexts[7] in terms of what will be here identified as the incipient moment of uncertainty which had its origins, above all, in geographical dynamics. It is the one that appears to emerge nowadays, because of the upheaval, that still has to find a geographical definition, of equilibrium and of global order (Kissinger, 2014), mainly in the geopolitical, economic and financial perspectives that were already felt at the end of the past century:

> a halo of uncertainty and hazard, of euphoria and depression, the quasi-cyclic oscillation between perceiving new virtues of peace. . .: this is what finally seems to be the destined result . . . in these few years that separate us from the end of the short century.
>
> (Veca, 1997, p. 227)

On the basis of which theoretical assumptions is it possible to compare these two, apparently very distant, historical moments?

Amongst others, there are three key topics which define, better than many others, the two epochal stages of the great change in society and territorial administration: *chaos*, *crisis* and *revolution*. All three terms denote – and are synonyms of – a moment of deep change that corresponds to the uncertainty that we are defining. This indefiniteness takes place in certain conditions, which, on an existential level (once again: singular or systemic), are reflected in a tendency of loss of certainties, in the political vagueness and in the fundamental *crisis* of a political system. In other words, if the three terms analysed in the previous paragraph (*insecurity, indecision, vagueness*) convey, through different paths, the conceptualization of *uncertainty*, a political response can be foreseen in the ideas of *chaos, crisis* and *revolution*.

First of all, *chaos*. What determines it? How can we recognize it? The lack – or the end – of stability and order, of any sort, that was guaranteed by given certainties, determines the moment of chaos. It is configured when certain univocal principles, capable of ensuring a personal or collective stability, cease to be. It corresponds to what Wallerstein (2011) and then Arrighi, from an economic point of view, defined as a "transition" (Arrighi, 1994, p. 77), that is to say, a shift from one recognized system within a certain hegemonic economic power to another.

The same dynamics could be witnessed in a more political, and more specifically geopolitical context.[8] The underlying concept, according to which a crisis that produces systemic chaos is witnessed, is that there is a passage from an order – basically recognizable due to the presence of a centre inside a system – to a phase of transition, in which the existential, political and economic references which animated the previous asset are lost: this is the meaning of transition from order to chaos, disorder.

The Modern Age European crisis was due, above all, to the disappearance of a feudal kind of political asset, which in spite of political turmoil and medieval centrifugal jolts had represented a systemic order vowed to universality and to the establishment of existential, political and social certainties. The beginning of Modernity, with its internal revolutions, the protestant movements, the erosion of hegemonic (spiritual, economic and political) centres and the conflictive stirrings, as well as the great geographical discoveries, led to the European *general crisis*.

In the post-Cold War global context, chaos was characterized by the absence of equilibrium, which had been previously guaranteed by the bipolar contrast – clearcut and certain in its ideological and political clash between the United States and the Soviet Union – which was not replaced by another global order.

Thanks to the maintenance of a constant tension of powers, the two countries had guaranteed systemic certainties, firstly within their regional spheres of influence and action, and subsequently in the whole world. When this global setup failed, certainties collapsed and propositions that could overcome the lack of the former order emerged, but they turned out to be unsuccessful or not farsighted enough (see Fukuyama, 1992). Consequently, there was a decisive moment of *general crisis*.

On further consideration, the terms *general crisis*, that differs from the one of *particular crises* due to its *extensive* nature (Colombo, 2014, p. 47), and *uncertainty* are evidently closely related to each other. They go hand in hand and are intimately interrelated: uncertainty originates from crisis, and crisis is substantiated in uncertainty. However, in its different interpretations, crisis can be imagined as a crossroad placed in front of man (Cesaretti, 2013). In this sense, that moment of separation and therefore, choice, always constitutes a substantial situation of uncertainty. In fact, "crisis generates uncertainty and it demands a decision – if anything, according to its original meaning, crisis *is* decision" (Colombo, 2014, p. 21).

This is also evident with regard to geographical arrangements and adjustments in the relations among States together with their spatial delimitations. Therefore, a general crisis always corresponds to a redefinition of borders. This in itself provides us with a prerequisite for strong bonds between the concepts of general crisis and geographical uncertainties: "crisis is both contraction of time and loss of borders; with the aggravating circumstance that the loss of borders accelerates the contraction of time" (Ibid., p. 47).

In a *general crisis*, which is also *systemic* as much as *extensive*, both the spatial and temporal spheres are involved in a progressive and ever more evident indeterminacy. In fact, the *critical* process foresees on the one hand, that time perception is reduced and, on the other, the dissolution of certainty of border settlements. The latter occurrence, in turn, affects the first fluctuation, contributing to create further crisis and uncertainty.

The European modern crisis – regarding identity, internal policy, reference values that were no longer universal or universally recognized and lived and perceived centrality – was a moment of uncertainty for Europe. It was predominantly originated by geographical dynamics connected to the discovery of a New World but also to revolutionary movements that formed the basis of the systemic transformations of the world.

The 17th century represented the peak of the *crisis* that crossed Europe during Modernity: a crisis that, above all, could be identified as an "identity crisis" compared to the past, and also cultural and philosophical one which had its origins in the internal religious, confessional and then political European conflicts. The critical-conflictual element characterizes Modernity: "*modernity is defined by crisis*, that is born of the uninterrupted conflict between the immanent, constructive, creative forces and the transcendent power aimed at restoring order" (Hardt & Negri, 2000, p. 76) and "this tension between ebb and flow, stagnation and progress, that characterises the period, will be . . . indicated with the term 'crisis'" (Kamen, 1971, p. 3).

Basically, it is the crisis that establishes modern European identity, in its geographical discoveries and the extension of Europe. Michael Hardt and Antonio Negri observed such ideas of association between Modernity and the concept of crisis in their book *Empire* (2000), which addresses the theme of globalization and its diachronic developments to the present day starting from the modern revolution crisis. It discusses the advent of Modernity, stating that it provoked a conflict that expressed itself in various forms of human knowledge and political action.

The European general crisis detected by Hardt and Negri in their book originated, therefore, from the great geographical discoveries that contributed to a state of generalized uncertainty, but it was also a crisis of European politics, derived from its internal multiplicity which contributed to establishing a specific identity in a multiplication of sovereign identities (see Tilly, 1994) that had to confront one another in the annulment of possible individual superiority in what would be defined as the invasion of balance of power (see Bull, 2002).

There is an aspect of divergence, of revolutionary crisis[9] in the aforementioned definition of general crisis, loss of borders and temporal contraction. We find ourselves intellectually, politically and morally in a new era, a new climate. As if, all of a sudden, a series of thunderstorms had blown into a single storm that had changed everything including the European temperature, for good. From the end of the 15th-century to the half way through the 16th-century, a single climate was experimented: the one of the Renaissance (Burckhardt, 1960). The years of change, of revolution, would come later in mid-17th century (Trevor-Roper, 1967, p. 46). Therefore, the Modern Age, with its European crisis, is seen as a moment of political, religious, cultural and geographical changes, corresponding to an actual revolution. According to Henry Kamen, "So profound are the changes of the post-Reformation era that many historians continue to look on the religious, economic and political events of the time as being revolutionary in character" (Kamen, 1971, p. 3). Indeed, also Hardt and Negri stated that "it all began with a revolution" (Hardt & Negri, 2000, p. 70).

Revolution and *crisis* fully characterize the advent of Modernity: "modernity's beginnings were revolutionary, and the old order was toppled by them", because a subjectification process was activated, coinciding with the crisis (especially of values) of the medieval system (Ibid., p. 74).

The *revolution* of Modernity was indeed sustained in the European *crisis*, by configuring a state of tension, of temporal and spatial loss, of vagueness towards the European political and social future (once again: uncertainty towards the future, towards destiny), that coincided with another kind of uncertainty, derived from the extension of the concept of Europe, of global spaces more in general and of borders between political entities of the period (Parker, 1980; Farinelli, 2008).

In fact, to better explain the process of systemic change, we can say with Hardt and Negri that the revolution took place in separate geographical directions (within and outside European borders) and in a subjective sense, in a process that led European men to get so far away from religious power and the metaphysical vision as to proclaim themselves "masters of their own lives, producers of cities and history, and inventors of heavens. They inherited a dualistic consciousness, a hierarchical vision of society, and a metaphysical idea of science" (Hardt & Negri, 2000, p. 70).[10]

The revolution started with Modernity (Goldstone, 1991) which strongly questioned the methodological, practical, cultural and representative implant proper of the Middle Ages, based on metaphysical and religious assumptions and supernatural references that could be defined as "certain", as it was considered metaphysical, and often dogmatic (see Harvey, 2006).

From a geographical perspective, that crisis originated with the discovery of the New World and, on the European side, by the internal dynamics of confessional nature within the Catholic Church which reflected the European conflict in the 16th-century: these two elements gave the European crisis its deepest character, which became coincidental with its geographic expansion on a global scale (Trevor-Roper, 1967, p. 1).

European 16th-century was marked by firstly religious and then political conflict and by cultural, scientific and geographical turmoil[11]: that same conflict involved every level of human action and knowledge, it did not stay within European borders but it was projected overseas, where European political projects had spread beyond the Pillars of Hercules, following the colonial and power fluxes of the European States towards the Americas:

> The internal conflict of European Modernity was also reflected simultaneously on a global scale as an external conflict. The development of Renaissance thought coincided both with the European discovery of the Americas and with the beginnings of European dominance over the rest of the world.
> (Hardt & Negri, 2000, p. 76)

In the 16th-century, the extension of European spaces was almost definitive, and it totalized the way of understanding the Old Continent and its own conceptual definition, of thought and identity, so much so that modern Europe itself would find its own outlet in the New World: "the arena of conflict among European States was widened to include lands and waters far beyond Europe's traditional limits, the pillars of Hercules" (Elliott, 1970, p. 79).

Hence, a *modus operandi* fully burst forth in all its vehemence, a way of conceiving the world and its political logics definable on the global scale, starting from that crisis that began precisely because of the extension of European borders: it is what Carl Schmitt would define as "global linear thinking" (2003, p. 87), which will be discussed later (Elden, 2010).

These two aspects are essential: European extension on the global scale and the maintenance of two fronts in the conflictual logic that defines European policy. No longer a single one, that enflamed the Old Continent with religious wars, but the opening of another battlefield, where internal European logics were projected and where new political, religious and cultural models could potentially be proposed.

It is possible to define such phase as a process from transcendence to immanence, producing secularization and a tendency to subjectivity. It is, moreover, a revolutionary process, because it turned the European geographical and political tendency to imperialism, universalism,[12] such as the imperialist medieval contest, into a contest of coexistence, of a pluralism of religious visions that were manifest in the complementarity of political entities (see Wight, 1977).

This new undefined and continuously changing set-up, subject to reformist pressures and to ideological impulses, determined by the Reformation and by the revolutionary currents typical of Modernity, could easily be included in the riverbed of modern uncertainty (Sloterdijk, 2013).

A further dilemma now arises, which is not the case to address here, but which remains of epochal importance and to which we will quickly respond using other authors' words: was it the reduction of all things to the mere human level to have consequences on all the rest, or vice versa?

According to Hardt and Negri, who have perfectly caught the spirit of those changes,

> the origins of European Modernity are often characterized as springing from a secularizing process that denied divine and transcendent authority over worldly affairs. That process was certainly important, but in our view it was

really only a symptom of the primary event of Modernity: the affirmation of the powers of *this* world, the discovery of the plane of immanence.

(Hardt & Negri, 2000, p. 71)

Hence, it can be more concisely affirmed that

the revolution of European Modernity ran into its Thermidor. In the struggle for hegemony over the paradigm of Modernity, victory went to the second mode and the forces of order that sought to neutralize the power of the revolution. Although it was not possible to go back to the way things were, it was nonetheless possible to re-establish ideologies of command and authority, and thus deploy a new transcendent power by playing on the anxiety and fear of the masses, their desire to reduce the uncertainty of life and increase security.

(Ibid., p. 75)

Here already, the clear idea of uncertainty of the modern condition can be seen. It originates concurrently with the affirmation of Modernity itself, with its conflicts and its *polemic* internal dynamics. An essential definition of Modernity intervenes, which has nothing to do with the war in its narrowest sense, but rather refers to a condition of internal conflict, of civil war.

The *crisis* was implemented by that "Thermidor" and the "civil war did not come to an end, but was absorbed within the concept of Modernity" (Ibid., pp. 75–76): it is also what Schmitt affirms when referring to the conflictual dynamics within the Old Continent, and it is what Federico Chabod emphasizes in regard to the origins of the idea of Europe, coinciding with the polemic and strongly conflictual reality of modern Europe (see Chabod, 2007).

Therefore, the first European Modernity is distinguished by its internally conflictual character and the achievement of a change of perspective and *forma mentis* that originated from geographical dynamics which configured a shift of centre, and by hegemonic transformations, mentioned by Arrighi and Silver, in the loss of medieval certainties and the new sociopolitical and geographical-political setup. It corresponded to a *crisis* of the old model, to the advent of modern *revolution*. It can be affirmed that during that historical time, "rebellions and revolutions shook regimes across the Eurasian continent" and determined its identity (Goldstone, 1991, p. 1).

Modern Europe *general crisis* coincided with a perception of temporal contraction and a redefinition of the borders in the *chaos* generated by new European and international political dynamics that would establish a more or less defined setup with the Peace of Augsburg (1555) and then with the Peace of Westphalia (1648). Moreover: *Modernity* itself finds its meaning in the concepts of *systemic chaos, revolution* and *crisis* that ontologically define it (Hardt & Negri, 2000).

Uncertainty and loss of a geographical centre

Surely the European crisis was also expressed in these internal conflictual dynamics, but where did it originate? When did it start configuring and why? The first step is to comprehend that, more than anything, it was a crisis which was originated from a geographical revolution,[13] otherwise all the rest would seem incomplete and only partially comprehensible.

By this, it is meant that Modern Age European crisis arose from the loss of certainties of its own centrality in international relations dynamics, because Middle Age Europe certainly was self-centred, and based its self-centeredness on mystical, and at the same time, symbolical bases which gave it a character of solidity and demonstrable evidence (Roux, 1985).

This was a vision proper to all human civilizations that tended to universalism, but what gave this unique character to European medieval civilization? Jean-Paul Roux's answer is immediate: its originality in this centripetal vision of the human soul derived from its "ignorance" and its exceptionality. Contrary to all the others who knew more, Europe believed that nothing, or almost nothing, existed outside its borders. Moreover, Europe itself affirmed the idea and conception of divine incarnation, giving the Old Continent a character of felt and experienced exceptionality, both in practical and spiritual terms (Ibid.).[14]

Europe was based on some certainties, which coincided with the spiritual and ecclesiastic truths, guaranteeing European men the safety of being part of the only accepted truth, the one given by the establishment of a sure and absolute centre. Those truths would correspond to dogmas, which as such are indisputable, and therefore spiritually superior to any other truth, even those scientifically provable: the same thing also happened in the geographical field, where European centrality had been established as unconfutable truth, one of the certainties around which the lives of men and society revolved.

Modernity was indeed mainly imposed by two historical events, which determined its revolutionary and critical significance: the voyages of discovery undertaken after Columbus' expedition – with the affirmation of European powers in American territories and elsewhere – and the protestant Reformation. The first extra-European development surely highlighted the Renaissance – it was a fundamental existential basis – but it also provoked a slow but progressive loss of European certainty: while the Reformation and the secularizing solicitations led to some revolutionary transformations in European *forma mentis*, and to emerging conflicts between European powers, it also caused a process of erosion of the catholic centre, which led to the collapse of previous values and the geographical-political setup. This second inter-European development travelled on a parallel track to the first (the discovery voyages) and was directed outside European borders: together they provoked the European crisis, which was, above all, identitary, but also geographical and geopolitical.

The Age of Discovery was, without doubt, a moment of systemic change for Europe, in the way of intending global relations and in geographical categories themselves, that were deeply mutated by the "New Geography" wave which was a result of the exploration voyages (Turco, 2010). It can be affirmed that the origins of modern uncertainty lays, most of all, in its geographical roots and in the revolutions that occurred in Europe after the discoveries that projected Europeans outside the Old Continent.

The oceanic routes became the highly dramatic theatres of international relations: here "the dominance of few countries over vast areas of Africa, America and Asia took shape and relationships between dominant countries developed" (Vallega, 1985, p. 55). The opening up of the western frontiers and the new markets in America meant a change of perspective from the Mediterranean to the Atlantic (Bailyn, 2005), and also a transformation in the philosophical approach which derived from the geographical knowledge field: as noted by Richard T. Rapp: "the change of axes of the European internal and international commerce was the result of the geographical discoveries that opened new horizons and opportunities for trades that became more accessible to the coastal nations in northern Europe" (Rapp, 1975, p. 500).

The spaces widened and narrowed at the same time, also thanks to the possibilities offered by an ever more efficient and extensive communication network between European States, and also between colonizer countries and conquered lands. In Schmitt's opinion, geographical discoveries and the circumnavigation of the Earth were at the same time "manifestations and consequences of transformations that lie deeper" (Schmitt, 2015, p. 57), of a change of men that was reflected in every field of knowledge and scientific speculation, from arts to philosophy, from literature to architecture.

Amongst others, modern cartography is also a vivid example of a natural process of osmosis when compared to the changes in geographical knowledge: no longer product, but origin of such incomparable revolution (see Farinelli, 1992, p. 72; Caraci, 2009, pp. 324–345). More generally, it can be said that for geography, the 16th-century

> marks the beginning of a meaningful transformation, meaning a progressive differentiation of what had been its preferential fields of study since ancient times: the first, also chronologically, the one of descriptive geography. . .; the second, the one developed by the Greeks, of astronomical-mathematic geography.
>
> (Caraci, 2009, pp. 337–338)

The meaningful change in the approach to geographical knowledge must be considered. According to Angelo Turco's words, in fact,

> Modernity marks the advent of a time in which geographers start to talk about a world where all things are in their place. Leading figures in the cartographic field established themselves in the scientific community in the second half of nineteenth century, they took on the task of describing the order of the Earth. Not without epistemological tempests, of course.
>
> (Turco, 2010, p. 277)

Here lies another match and another dilemma to be solved, found in the very expression used by Turco who defined this representation as more precise in cognitive terms.

The question to be asked in this specific case is: how can the acquisition of cartographic, territorial representation and scientificity go hand in hand with what is being defined as the geography of uncertainty, in the continuous dissolving of territorial elements, that should yet assume a crucial role in the exercise of public power? We will attempt to answer these questions in the chapter regarding cartography, its tragedy and the paradoxical uncertainty that was verified in European Modernity.

It is a paradox, indeed, because with Modernity more reliable information was acquired compared to the past, because with time, the scientific knowledge of the world became more stable and because we entered the domain of what we could define as the "certainty of representing", right at the time of the "image of the world" (Farinelli, 1992, p. 55).

The "epistemological tempests"[15] we referred to sprung from the knowledge of the globality of the world and from the "loss of European centrality". They represented a real shock to cultural and scientific categories and concepts pertaining to geography and the related cartographic representation.

Another question necessarily arises: was this proof of European centrality, or was it a reaction to what was beginning to emerge as the loss of centrality of the European continent itself?

In this regard, several interpretative formulations have been offered on Europe's role and on the perception that Europeans had of it. While it can be thought that this shift of European barycentre produced a loss of a typical eurocentric logic pertaining to the Middle Ages,[16] some authors such as Samir Amin (2010) foresaw, at that moment of European opening to global spaces coinciding with the Early Modernity, a cultural and political crystallization of eurocentrism. This meant that Europeans, knowing that there were other spaces to conquer, tried to further claim their centrality, a centrality that was attempting to extend the European internal dynamics to other continental contexts.

It is as if it were an overflow from European borders, to reproject Europe as the centre of world destiny, through its action of discovery and conquest, of giving new strength to eurocentrism, that inevitably would have been eroded in time, by the great exploration voyages.[17] During the European Renaissance, the principle of European centrality held true and was to be established on nationalistic levels or on new found territories. Those same explorative voyages, that projected Europe outside itself, would represent the seed of the future and of irrefutable loss of its centrality, or of its progressive marginality, as already understood by the Europeans in a cartographic sense.

From a representative point of view, European cartographers would try to put Europe in the centre in any case, by means of that phenomenon of national and self-centred affirmation, but that was not enough to maintain the global barycentre. The change was obvious when compared to the *mappaemundi* in which Europe, and the Mediterranean in particular, was central in the dynamics and in the destinies of the world.

According to Hardt and Negri, "eurocentrism was born as a reaction to the potentiality of a newfound human equality; it was the counterrevolution on a global scale" (Hardt & Negri, 2000, p. 77). The authors' position also confirms the data that clearly emerges: the inevitable loss of European centrality arising from the discovery of America.

It is surely true that the discovery of a New World contributed to crystalizing eurocentrism, as affirmed by Amin, but it was a reactionary movement, more than true awareness. A reaction due above all, to uncertainty that the projection towards an unknown continent which had never been explored by Europeans, had contributed to creating and was substantialized in the full awareness of a globality of the world, that led to the loss of that *European egocentrism* recalled by Roux.

Enormous geographical uncertainty remains in both visions, the one that established a new European centrality with the discovery of the New World and the one which saw its substantial loss because the first case consisted in a reaction and the second, in the acceptance of a widening scale, inevitable and ineliminable at the moment of the discovery.

Times and ways of modern uncertainty

Similar dynamics, relative to the loss of fulcrums around which a global balance was established during the Cold War, came to pass after the fall of the Berlin Wall, when the conflict deriving from national belongings widely and dramatically resurfaced – in an abruption of geopolitical chaos and general crisis – multiplying the centralities, with the unescapable consequence of the systemic breakdown of the previous order (Brecher & Wilkenfeld, 1989).

Indeed, according to a certain interpretation, the current situation of global chaos[18] and uncertainty, as noted by several authors and as will be seen more specifically later, seems to be a further focal moment, or maybe the last step, of a long-term process started with the Early Modernity and with the discovery of America by Cristopher Columbus, which saw its first crucial moment in 17th century (Arrighi & Silver, 2001; Rosenau, 1990).

This was the idea expressed by many scholars, when discussing the last globalization processes, which lead to the end of the bipolar world and the closure of a historical moment which had been held together by the separation between two blocks and its relative global order (Galli, 2001).

In a long-term consideration on politics, geography, economy and, more in general, on global geopolitical dynamics, other authors have compared the nowadays scenario to the one of European Modernity, in the period following the Age of Discovery.

The current condition, in fact, could be interpreted – from the point of view of the fall of certainties – as analogous to the crisis that shocked Europe during Modernity. Similarly, the same erosive process of previous certainty systems took place with the "geopolitical seism of Eighty-nine", subsequent to the fall of the bipolar system, in which the "collapse of large reserves of stable beliefs in durability" was lived (Veca, 1997, p. 227) and was, therefore, tied to the safety system of the Cold War.

18 *For a definition of uncertainty*

Quoting Foucault, a parallelism relative to the shift from the feudal system to the modern one can be proposed and, *mutatis mutandis*, from the bipolar one the setup following the Cold War:

> [I]n a completely general, rough, and therefore inexact way, we could reconstruct the major forms, the major economies of power in the following way: first, the state of justice, born in a feudal type of territoriality and broadly corresponding to a society of customary and written law, with a whole interplay of commitments and litigations; second, the administrative state that corresponds to a society of regulations and disciplines; and finally, a state of government that is no longer essentially defined by its territoriality, by the surface occupied, but by a mass: the mass of the population, with its volume, its density, and, for sure, the territory it covers, but which is, in a way, only one of its components.
>
> (Foucault, 2009, pp. 109–110)

In these historical passages, the germs of the current geography of global uncertainty are detected: geography that takes shape in the lack of a global geopolitical centrality, in the increasing porosity of inter-State borders, also marked by the general *indecision* and *insecurity*, as well as symptom of the affirmation of *vagueness* and of a *general crisis* that involves multiple compartments of human thinking and action.

It is now useful to understand which directions uncertainty follows today, or better how such a topic has been analysed by other disciplines, also in order to attempt to see different authors' given definitions.

Many scholars have indeed spoken about uncertainty from multiple perspectives, trying to explain the current condition of man in society, in financial economy and from a philosophical point of view. These implications of uncertainty will help us better understand what is meant by the proposed idea of geography of uncertainty and if it is possible to link the first globalization of the Modern Age to the current international geopolitical structure.

Notes

1 A more extensive review of the geographical debate will be proposed in the next chapter.
2 See Ricci, 2018b; 2021.
3 See www.newworldencyclopedia.org/entry/Uncertainty.
4 In this regard, please refer to the very interesting definition of *crisis* given by Colombo, 2014.
5 This is a concept widely analysed from the point of view of the geographical representation and the use of GIS. See, for example, Fisher, 2000.
6 The same comparison between the post-Cold War system and the one that found its asset in 16th-century Europe is made by Arrighi and Silver, claiming that "in his view, the parameters that have framed action in the international system are being transformed so fundamentally today 'as to bring about the first turbulence in world politics since comparable shifts culminated in the Treaty of Westphalia in 1648'" (Arrighi & Silver, 2001, p. 257).

For a definition of uncertainty 19

7 This crucial matter will be analysed further, where an analysis of the Modern Age as a period of revolution, crisis and as a revelation of the moment of uncertainty will be attempted.
8 It is not casual that the same Arrighi and Silver talk about geopolitical transitions, where it is the modern economic dynamic itself to establish the hegemonic assets also from a political point of view.
9 In this regard, please refer to Rosa, 1977: in this collection of essays, the concept of crisis is widely brought to light through the work of various and authoritative historians. See also D'Eramo, 1980; Le Roy Ladurie, 1980; Béjin & Morin, 1980.
10 According to Trevor-Roper, "it took more than a century for Europe to emerge from the crisis originated within its own social structures" (Trevor-Roper, 1967, p. 6).
11 Please refer to Trevor-Roper (1967), chapter II, entitled *The General Crisis of the Seventeenth Century*, where the general revolution which happened during Modernity is well explained: "it is an interesting but undeniable fact that the most advanced scientists of the early sixteenth century also included the most learned and literal students of biblical mathematics; and in their hands science and religion converged to pinpoint the dissolution of society, the end of the world between 1640 and 1660" (Trevor-Roper, 1967, p. 44).
12 Concerning the imperialist logic, see Münkler, 2007.
13 Chabod perfectly affirmed this idea when he wrote that the great geographical discoveries, and particularly the discovery of America, indeed affected economic life, but also and not less deeply, the European spiritual life (Chabod, 2007, p. 58).
14 This same opinion was also expressed by Brunner, when he affirms that Middle Ages' West is considered to be coincident with the Cristian and Roman Church (Brunner, 1978, p. 37)
15 In fact, between Modernity tempests and the current ones, there is a strict correlation: if during Early Modernity – as stated by Angelo Turco – "epistemological tempests" were observed (2010, p. 277), the same dynamics can be foreseen in current international political relations on a European scale (see also Colombo, 2014, p. 30).
16 A shift in the international trade centre from the Mediterranean, that had been the point of convergence of medieval traffic, to the Atlantic happened, and with this the decay of the economic fortune of the great Italian republics (Chabod, 2007, p. 58).
17 According to this view, "the discovery and settlement of the New World were incorporated into an essentially Europocentric conception of history, where they were depicted as part of that epic process by which the Renaissance European first became conscious of the world and of man, and then by degrees imposed his own dominion over the newly-discovered races of a newly-discovered world" (Elliott, 1970, p. 3).
18 There is a wide bibliography in this regard. Among the main references, please see: Amin et al., 1982; Arrighi & Silver, 2001; Barnes, 1999; Bauman, 2000; Berman, 1998; Hannay, 2008; Horsman & Marshall, 1994; Jowitt, 1992; Khanna, 2011; Rosenau, 1990; Thrift, 1989; 1992.

2 The society of uncertainty

Uncertainty in geography and its geographical representation

The concept of "uncertainty" has been investigated over time from many points of view and from time to time has acquired specific meaning depending on the disciplinary context of analysis. In the geographical field, the topic of *uncertainty* has rarely been connected to the context of knowledge of the world and the structure of global relationships. Few and disparate authors have addressed the topic, and often from completely different perspectives, while remaining within the scope of geography. Some of them – mostly in economic geography and in border studies, as we will see – have spoken of single and specific topics, inserting them into an *uncertain* global scenario (see Yeung, 1998; Silvey, 2010). In other words, the uncertainty in the world has been seen only as a secondary point of the question, or just as a root of other problems to be studied.

Nigel Thrift, as a geographer who connected history and geography, pointed out the crescent uncertainty in capitalism dynamics of the present world scenario. In his reflection, capitalism is strictly linked to an uncertain context:

> [T]his book is therefore, on the one hand, a history and geography of the nearly present, battered this way and that by the power of events. But it is also, on the other hand, an attempt to show how that very uncertainty is increasingly being taken up and worked with by capitalism in ways which are productive of new kinds of aggregation and ordering which have most decidedly not been present before and which need pointing to, since on them are being built new means of reproducing an order which many would argue has life-denying characteristics.
>
> (Thrift, 2005, p. VI)

A strong debate regarded the use of data within geographic information system (GIS) technologies and how uncertainty could be mastered in geographical representation systems. There has been a focus on this topic since the late 1990s, when the question of uncertainty arose in certain studies on GIS usages and on the spatial dimension (Thill & Sui, 1993; Windholz, 2001; Agumya & Hunter, 2002;

DOI: 10.4324/9781003394204-2

Foody & Atkinson, 2002; Zhang & Goodchild, 2002; MacEachren et al., 2005), in spatial databases (Morrey, 2007) and for the studies on remote sensing systems, as well reported by Peter M. Atkinson and Giles M. Foody (2002). As outlined by WenZhong Shi, GIS are deeply and strictly related to uncertainty. In his opinion:

> [T]hree areas are defined as the major principles of geographic information science: (a) space and time in GIS, (b) uncertainty and data quality, and (c) spatial analysis. In fact, uncertainty modelling covers these three areas, together with the most obvious one, data quality.
>
> (Shi, 2008, p. 38)

At the same time, uncertainty has been viewed as an umbrella term covering different areas such as "incomplete knowledge, inaccurate knowledge, imprecise knowledge, fuzzy knowledge, disputed knowledge, ambiguous knowledge, impossible knowledge" (Fusco et al., 2017, p. 2262). Yongwan Chun et al. (2019) have also focused their attention on uncertainty in GIS and representation in sessions dedicated to this topic in an American Association of Geographers (AAG) conference, highlighting, in a sort of review, the main uncertainty topics related to geography and geographical representations (see also Kwan, 2012). Helen Couclelis stated that "in GIS science, research on handling uncertainty, error, inaccuracy, imprecision, ambiguity, vagueness, and related issues has focused for the most part, on problems with spatial data and their direct products, typical representations of spatial objects or fields" (Couclelis, 2003, p. 166).

From a similar perspective, regarding mapping for historical purposes, Brandon S. Plewe outlined that "a major difficulty in implementing a permanent historical GIS is that historical geographic information is inherently uncertain or imperfect. Uncertainty is common in most types of GIS information" (Plewe, 2003, p. 319) and stated at the end that "some of the basic uncertainty management strategies presented here may have wide usefulness in GIS, such as attempting to store primary data in permanent databases rather than derivations" (Ibid., p. 333). For Ashley Morris, "*Uncertainty* simply means that of which we are not certain; that which is not known. Uncertainty can be divided into ambiguity and vagueness" (Morris, 2007, p. 80).

The Congress of American Geographers, in 2018, focused on uncertainty in the related special issue of "Geoforum" which was dedicated to the topic of *Geographies of uncertainty*. In particular, the editors explicated that theirs was a qualitative approach to uncertainty in geography:

> [W]e explore how uncertainty is often produced through, and temporarily stabilized by, key geographic axes such as time, space, scale, and regimes of environmental governance. Attending to the ways that uncertainty is experienced as a spatiotemporal condition, and how it frequently compounds across scales of knowledge production.
>
> (Senanayake & King, 2021, p. 130)

The authors widely reflected on how uncertainty has been dealt with in the geographical field, having socio-spatial implications and connections with other disciplines (see Scoones, 2019; Shattuck, 2020), how uncertainty can produce possibilities, "how uncertainty can restructure social and ecological relationships", as well as "how uncertainty is reconfigured through human-non-human dialectics" (Senanayake & King, 2021, p. 130). At the same time, they pointed out how uncertainty has been interpreted by geographers as a pathway to reconfigure the relations among human beings and environmental politics.

Peter Fisher defined geographical thinking in terms of "vagueness": he explained that it "is endemic in geographical thinking, and in geographical information" because "vagueness is an inherent property of geographical data" (Fisher, 2000, p. 8). More recently, a congress focusing on topics including Early Modern history and mapmaking was organized at the Royal Netherlands Institute in Rome, entitled *Mapping Uncertainty. Early Modern Global Cartography, 21st Century Discussions*.

Since geographical research has given no particular contribution to this specific concept, especially in relation to globalization dynamics, the aim of the following considerations hopes to provide a key – surely not an exhaustive one – to interpretations of the idea of uncertainty in existing literature, also included in other disciplines.

The society of uncertainty

In sociological and anthropological areas, many studies have pointed out the condition of looseness of men in the Modern Era, connected with a renovation of places and in relationship with territory.[1] In the field of modern society studies,[2] at the beginning of the 20th century, many theories were elaborated on city changes, seen – with the coming of mass industrialization and structural and institutional urban modifications – as a place where human fears were concentrated and no longer linked to an organic and communal context of family relationships, but to one of "Society" ("*Gesellschaft*") (Tönnies, 1922), in which economic dynamics acted as a pivot of social relations. In urban areas, man is alienated from his social context, considered a "jungle" characterized by social relations based only on economic exchanges. All this contributes to the perception of confusion and estrangement in the citizen, turning him into a foreigner in his own country (see Weber, 1978).

The School of Chicago (with Robert E. Park, Florian Znaniecki, Louis Wirth, Ernest Burgess and many others) offers a fundamental and analytical contribution, mainly focused on the evolution of urban contexts at that time, and more specifically, on the studies of the 1920s after the Chicago fire of 1871 and the changes it determined both in human behaviour and in urban contexts.[3]

Many of the authors quoted underlined the great uncertainty of the future in modern cities, and how social relations were structured there. More recently, also considering the theories put forth by the School of Chicago, numerous other researchers have outlined the state of strong uncertainty concerning the future, viewing it as a society of risk (Beck, 1992; Luhmann, 1993; Zinn, 2008), mostly after world changes in the 1980s and after the fall of the Berlin Wall and the related global geographical-political transformations. Ulrich Beck targeted the signs of deep society insecurity that will emerge after the Cold War which were more relevant in economic flows rather than on a social level.

In his work *The Risk Society* (Beck, 1992), he identified in a pessimistic view the risks resulting from the capitalistic global system entirely projected on economic profit, where the imperatives are profit and economic productiveness, with intensive and unbearable rhythms, where there is no space for product safety nor safety in the workplace, and neither for the human condition.[4] In the world depicted by Beck, in substance, becomes global and risk is the only possible conclusion and, quoting his very expressions, "the uncertainty arises" (Ibid., p. 180) and "the uncertainty grows" globally (Ibid., p. 171).

Over the past 20 years, Zygmunt Bauman has discussed, in different works, theses on "liquid Modernity" (Bauman, 2000), on the destructive power of world economy on human identity within a global context, strongly asserted after the fall of the Berlin Wall. In the changed global context he describes, in the "society of uncertainty" where human beings are alienated from the surrounding society, they move from the condition of "pilgrim", obsessed by the building of their own identity in the modern world, to the one of "stroller" ("flâneur"[5]), "vagabond", "tourist" or "player", symbols of man in post-modern life (Bauman, 1995, pp. 88–99). In Bauman's opinion, we are living "the uncertainty and instability of all things worldly" (Bauman, 2000, p. 154) as well as "the spectre of uncertainty is thus exorcized through regimentation. Certainty is restored by forces external to the individual – from outside. In the last account, the modern cure for uncertainty boiled down to curtailing the realm of choice" (Bauman, 1995, p. 108). In his analysis, Bauman seems to share Beck's opinion on the ideas of *insecurity*, *risk* and *unsafety* which are dominant in actual working contexts. To this purpose, Bauman outlines that

> the phenomenon which all these concepts try to grasp and articulate is the combined experience of *insecurity* (of position, entitlements and livelihood), of *uncertainty* (as to their continuation and future stability) and of *unsafety* (of one's body, one's self and their extensions: possessions, neighbourhood, community)
>
> (Ibid., p. 161)

Speaking of the actual work organization, concepts of disunity and related chaos of the world are united. These will be analysed in the next pages: "it is such a fluid form of assembly that best fits their perception of the surrounding world as 'multiple, complex, and fast moving, and therefore "ambiguous", "fuzzy", and "plastic", uncertain, paradoxical, even chaotic'" (Bauman, 2000, p. 154). And, continuing with the idea of anxiety and uncertainty, he explains that:

> the question of objectives is once more thrown wide open and bound to become the cause of endless agony and much hesitation, to sap confidence and generate the unnerving feeling of unmitigated uncertainty and therefore also

the state of perpetual anxiety. In the words of Gerhard Schulze, this is a new type of uncertainty: "not knowing the ends instead of the traditional uncertainty of not knowing the means".

(Ibid., pp. 60–61)

He also connects the idea of uncertainty to the key topic of the individualization of the world, which will also be identified as a turning point in the history of uncertainty in this book.[6]

This kind of reflection was also elaborated by Marshall Berman, who wrote the book *All That Is Solid Melts into Air: The Experience of Modernity* (1988) and included Marx's idea of uncertainty in the modern human condition in the title. The study focuses on and around urban sociology, stressing the idea of fluidity and uncertainty which dominate modern cities and modern environments. In a very clear passage, Berman underlines his reference to Marx, when he quotes him on the bourgeois condition and the differences with the past:

the bourgeoisie cannot exist without constantly revolutionizing the instruments of production, and with them the relations of production, and with them all the relations of society . . . Constant revolutionizing of production, uninterrupted disturbance of all social relations, everlasting uncertainty and agitation, distinguish the bourgeois epoch from all earlier ones.

(Berman, 1988, p. 21)

In Berman's opinion, indeed, "this is probably the definitive vision of the modern environment, that environment which has brought forth an amazing plenitude of modernist movements, from Marx's time to our own" (Ibid., p. 21). Moreover, this situation "leaves us all in strange and paradoxical positions. Our lives are controlled by a ruling class with vested interests not merely in change but in crisis and chaos" (Ibid., p. 95). He even refers to Friedrich Nietzsche and to the tragic condition of mankind after the "death of God", which took with it certainties that were no longer present: "it found itself in the midst of a great absence and emptiness of values and yet, at the same time, a remarkable abundance of possibilities" (Ibid., p. 21).

The words he uses could also refer to the human condition of five centuries ago, in some way:

[T]he glory of modern energy and dynamism, the ravages of modern disintegration and nihilism, the strange intimacy between them; the sense of being caught in a vortex where all facts and values are whirled, exploded, decomposed, recombined; a basic uncertainty about what is basic, what is valuable, even what is real; a flaring up of the most radical hopes in the midst of their radical negations.

(Ibid., p. 121)

The idea of actual human uncertainty seems to be well expressed also by Walter Benjamin (1966) with the image of an "Angel of History" "who is blown away by the storm of catastrophic history into an uncertain future and whose gaze wants to awaken the broken shards (of the past)" (Goebel, 2009, p. 3). This conceptualization of catastrophic history (Benjamin, 1977) will be then used in comparison to the idea of a "tragedy of modern cartography".

Uncertainty in the international political system

The global scale leads us to consider not only the urban possibilities stemming from world disorder, the relationship between local communities and those of immigrants or new classes configured in recent decades or in the links between migration flows and the more general implications of uncertainty, but also what happens to international relations and financial economy and the more or less immediate impact on relative territories. It is appropriate, therefore, to present a brief overview of the investigations carried out on international relations dynamic fluidity and on current geopolitics, trying to highlight the aspects that are strongly linked to the idea of uncertainty in political geography.[7]

For some scholars, the geographical and political "disorder" of recent years derives from economic world disorder due to the end of the bipolar system (Barnes, 1999; Hobsbawm, 1995; Rosenau, 1990; Wallerstein, 1995) and recent financial crises: they seem to be deeply reciprocally connected and both contribute to the political geography of uncertainty. In 2001, Giacomo Marramao noticed that the political emptiness produced by the decline of Nation States and of the network of international agreements, which occurred during the Modern Age, brought instability to light in the economical-merchant field, as well in the law systems and in international equilibrium (Bolaffi & Marramao, 2001, p. 27).

As Arrighi and Silver clearly stated when considering international relations in the last years, "a sea change of major proportions is taking place in the historical social system forming the modern world, creating a widespread sense of uncertainty about the present and foreseeable future" (Arrighi & Silver, 2001, p. 257).

Taking up the idea of multi-polarity, which seems to be one of the causes of modern uncertainty caused by the end of the bipolar system and American "hegemony" (Colombo, 2022), concepts such as "Fragmentation and System" (Bolaffi & Marramao, 2001) and "Disunity of the world" (Colombo, 2010; Khanna, 2011) started to be introduced through the analysis of the globalization process and its structural evolutions, presenting the idea of a globalization that is living a new phase.

Moreover: no longer is there a conviction of a stronger globalization based on the ever more sophisticated systems of mass communication (Appadurai, 2012), able to instantly connect different parts of the world; no longer is *sic et simpliciter* the certainty in a neo-liberal economic system that has a world scale, able to corroborate the idea of globalization (Friedman, 2005), but a new phase of globalization is appearing which started after the fall of the Berlin Wall and continued in

September 2001 with the attack on the Twin Towers and the economic crisis of 2007–2008 (Colombo, 2010; 2022).

This process is referable to the changes of the international political-economic structure, due both to the conduction of a "global war on terror" which has shown its strategic limits (Sheppard & Leitner, 2010; Smith, 2005; Benigno & Scuccimarra, 2007), and to the global economic crisis.

In this regard, it is perhaps useful to consider the fact that direct interconnection from the economic and social point of view between the different parts of "Western" economy seems undeniable, even if in the past few years, the concept of globalization has been seen in different lights.

This situation seems to have fully manifested itself precisely at the outbreak of the recent economic crisis, which, despite having covered a specific financial area and having had his driving force in some States of the United States, beginning at local levels to then spread nationally (Aalbers, 2009) and internationally (Agnew, 2009), affecting, almost immediately, other parts of the world, not least Europe and Italy.

On the one hand, the economic crisis of the last years has contributed to the rise of a widespread uncertainty, and on the other hand, that crisis can be seen as just another point of globalization of the past 30 years.[8]

Indeed, in the aftermath of the fall of the Berlin Wall (Hannay, 2008; Minc, 1994), the idea of global chaos and disorder fully emerged, due to the extreme fragmentation of a world that had lost any defined and stable structure in the bipolar division of the world of international relations. As Ignacio Ramonet pointed out from a geopolitical perspective, "on the eve of the Third Millennium, everyone can tell you that uncertainty has become the only certainty" (Ramonet, 1998, p. 13).

After a first moment of extreme optimism,[9] the concept of a very chaotic, unstable, uncertain global order began to take place. Some scholars found the cause in the transition from the emergence of distinct nationalisms to a return to tribalism, with its many elements of disorder.[10]

A very similar idea concerning tribalism was expressed by Salvatore Veca, when he wrote: "The oscillation between universalism and tribalism is, as usual, rather intense" (Veca, 1997, p. 226). In a different light, the same concept was also proposed by Kaplan, who, when speaking about the rise of "tribalism", proposed the idea of the "Revenge of Geography" (Kaplan, 2012), referring to the fall of the political and cultural world of the Cold War, highlighting the necessary reconsideration of the geographical factor in international relations,[11] in a context where the ideological and cultural factors had basically disappeared, as happened at the very beginning of the 1990s and as has been noted also in later stages of recent globalization, including the global wars on terror and the destabilization experienced in different geopolitical contexts, from Eurasia to the Middle East. John Agnew, from his point of view, elaborated the concept of the "territorial trap" (1994)[12]: the idea of that work was explicitly

> to identify and describe the geographical assumptions that have led international relations theory into the "territorial trap". To this end, the first section offers a short discussion of space and spatiality. A second section provides

a review of the position taken on the territorial state in the "mainstream" of international relations. The specific geographical assumptions that underpin the conventional representation are then examined. These are held to define a "territorial trap" for the field as a whole.

(Agnew, 1994, p. 54)

Alain Minc identified post-Cold War world reality with the conception of "new Middle Ages", which, however, did not provide any "regulatory" mechanism capable of bringing order: "neither the Cold War nor international alliances nor local imperialisms, work. The new Middle Ages are therefore aleatory, uncertain, evanescent" (Minc, 1994, p. 8).

The French author, in his certainly pessimistic essay on history and on the future of Europe,[13] identifies an element that appears to be very useful in understanding this discussion, which is an almost inextricable paradox, a seemingly incomprehensible contradiction: how is it possible that the extreme consequences of a world based on the certainties of science, on the "certainty of representation" and also on international relations produce, in the long run, a system defined as "new Middle Ages", which is totally uncertain and based on uncertainty? The problem is expressed by Minc in even clearer terms:

"for Cartesians who were used to transparent structures and to a seemingly rational world order, how difficult is it to understand such a fleeting reality, made up of an absence of incomplete power, a shattered economic network and a society that escapes more classic canons!", which have led to a situation of "absolute vagueness of the future".

(Ibid., p. 17)

Minc speaks of a clear, transparent Cartesian – we could say "certain" – structure of thought. The issue will be addressed more explicitly – and hopefully exhaustively – later in this work. However, it is appropriate to anticipate that the contradiction is, in fact, only apparent. It seems to be precisely the understanding of world and science, as formulated by Descartes, to have substantially contributed to modern uncertainty, becoming the "sublimation" of modern uncertainty, the conclusion of a long process of affirmations of uncertainty which had already appeared in modern, social political thought, both in the conception of the world and in relations among States, which is of more interest here.

The geopolitical reality of 1993 appeared to Minc as geographically uncertain, especially in the Old Continent: "fluidity dominates Europe, since its economic, political and strategic maps no longer coincide" (Ibid., p. 34): in such a situation,[14] the perspective was very negative and defeatist, similar to – as Minc affirms – a new Middle Age: in the worst scenario, Europe will be the continent of chaos; in the best scenario, the continent of complexity (Ibid., p. 37).

With the end of the Cold War – states Minc – which ensured certainty on the maps he referred to, "we are now returning to a normal state with its three diverging maps increasingly carrying dangers and risks in their junctions" (Ibid., p. 34).

That period, with its two opposing and ideologically rooted worlds, almost eternally projected into the future, seemed to guarantee a fixed order (Hobsbawm, 1995). The world, after the Cold War, for many authors was indeed "orphan of the certainties of bipolarity" (Minca & Bialasiewicz, 2004, p. 235).

Also Parag Khanna (2011), for his part, explains the current diplomatic world situation in terms of a new Middle Age. He stretches this idea and links the "new Middle Ages" to the same uncertainty when he says that "the new Middle Ages also need to be a permanent purgatory of uncertainty" (Khanna, 2011, p. 21), because we live in a world dominated by chaos:

> if there is such a hidden plan, it is unlikely to emerge during our current period of global uncertainty – even if now is when we need it most. Instead we are headed into the "turbulent teens", a decade (or more) of confusion, disorder, and tension.
>
> (Ibid., p. 208)

The comparison with the Middle Ages, proposed both by Minc and Khanna, is also shared by other authors: "*Neomedievalism* has been used to describe some of the aspects of the contemporary 'war on terror', globalization, or in terms of political geography the changes of the European Union" (Elden, 2013, p. 100). As far as the war on terror is concerned, Stuart Elden quotes Bruce Holsinger (2007), by using that suggestive idea of a new Middle Age applied to globalization. This conceptualization of *neomedievalism* was used to describe current international relations or affairs, even if in some way it radically differs from the here developed idea of uncertainty connected with Early Modernity (as reported by many authors already quoted).

The comparison with the Middle Ages (see Cerny, 1998; Elden, 2013),[15] as argued even later, is debatable, because if it is true that the Middle Ages represented a multitude of political entities, they were recognized in a centre – which was at the same time symbolic, political, cultural, economic and so on – as explained by Arrighi (1994) and Brunner (1978). Minc admits that the uncertain situation is precisely due to the absence of a centre, to the condition of a "world without a centre" (Minc, 1994, chapter II).

According to Otto Brunner, multiplicity in the Middle Ages had completely different characteristics from those of the Modern Age:

> [C]haracteristics of the western medieval times are, from one side, the unity of the church and, from the other, the persistent multiplicity of States, two elements which have deeply determined their internal dialectic of the system. Nevertheless, even in the presence of several States, the world of the Christian States differs from the surrounding world because it is also linked by a common law of peoples.
>
> (Brunner, 1978, p. 38)

The same idea was expressed by Arrighi, when he stated, bringing forth other authors' theses and pointing out the geographical dimension, that "the medieval system of rule consisted of chains of lord-vassal relationships, based on an amalgam of conditional were geographically interwoven and stratified, and plural allegiances, asymmetrical suzerainties and anomalous enclaves abounded", and indeed "finally, this system of rule was 'legitimated by, common bodies of law, religion, and custom that expressed inclusive natural rights pertaining to the social totality formed by the constituent units'" (Arrighi, 1994, p. 31).

Apart from this relevant difference concerning *neomedievalist* conceptions, the analyses of James Anderson (1996), Philip G. Cerny (1998), Holsinger (2007), Khanna (2011) and Minc (1994) seem, in any case, to fit in very well with the thesis here proposed, mostly concerned with current international relations as deeply dominated by geopolitical uncertainty. All the statements here quoted fit in with the post-Cold War context and determine an "economic map which goes beyond frontiers, States, limits", and "do not define a political space" because "several frontiers seem uncertain" and "in the long term, no one appears eternal" (Minc, 1994, p. 35). These ideas seem to be more applicable to Early Modern Times, when the Nation States were being formed against imperial principles, rather than to the new Middle Ages.

In the past few years, even some newspapers and journalists concerned with global, political and economic issues have highlighted the idea of a substantial uncertainty in international relations as a result of the bipolar confrontation: Guido Rossi (2014) has revealed these aspects, also by contrasting them to that Hegelian idea expressed by Francis Fukuyama (1992) immediately after the Cold War, following the end of these tensions due to the assumed absence of opposition against the overwhelming power of the United States in the early 1990s.

These considerations have given way, over time, to an awareness of the impossibility of establishing an order in international relations, especially – this is added here – as a result of two crucial events that have strongly questioned the leadership of the United States and, with it, the possibility of an order established by them: the terrorist attacks on the Twin Towers in 2001 and, a few years later, the economic and financial crisis started in the United States, that firstly destabilized its economy and then moved to a European context and not only, with different consequences and effects globally.

Hence, the inevitable conclusion of disorder, chaos and uncertainty emerged about the future, and it replaced the idea of a future of perpetual prosperity precisely dictated by the inevitable unipolarity and by an order established by the United States (Fukuyama, 1992). This is essentially the same concept expressed by the Italian philosopher, Remo Bodei, in an article entitled *Thinking the future, or the global uncertainty*, where he explains that "the decline of great collective expectations, that until a decade ago (when the world was still divided in two blocks) guided, albeit ideologically, billions of men, tends to lead to a privatization of the future itself" (Bodei, 2010, p. 2).

The author, in a more meaningful and effective way, had then expressed the idea of loss of certainties – especially in relation to the future – after the distancing from deep cultural, ideological and "religious" roots. He compared the current context to that of Tocqueville, in around 1840, and highlighted the major importance of present time and contingency, compared to the future, in the decadence of "eternal" principles.

Bodei, in his reflection on current times, considers that

> the contraction of expectations, purely in its physical existence, immerses the individual in the irredeemable time of transience, forcing him to process the loss caused by having to transplant the roots of his ego from the solid and unchanging ground of the afterlife or from the epochal times of History to the friable and fleeting soil of his own body.
>
> (Ibid.)

The same path was trodden by the rise of Modernity, when the world of medieval certainties disappeared: "traditional societies had quite effective structures, able both to compensate men from the possible disadvantages of their condition, and to justify hierarchies" (Ibid.), also, through the promise of a reward after death, which was a certainty for human life.

This process of loss of certainties, that Bodei connects to the decay of "traditional societies" or those based on strong ideologies, certainly has its effects on the quality of individual lives and on the design of the future, but it can be also seen in other contexts, citing Condorcet and the calculation of probabilities.

Bodei then moves to the sphere of global issues and considers all the components and the social and political factors in the current world:

> [T]he impressive development of techniques and scientific knowledge, the volatility of financial markets, the historical situation in which the great civilizations of the Earth still do not sufficiently recognize their peculiar principles, the bifurcation of centripetal processes of globalization and centrifugal processes of insulation.
>
> (Ibid.)

He wonders how much they can allow reliability of the future, the possibility of its prediction. His answer is clear: on the local, narrower scale, some answers can be provided in an almost certain way, but moving onto a global scale, degrees of arbitrariness and uncertainty appear, that are "easily measurable through the gap between the present future and the future present" (Ibid.).

In this, Bodei also necessarily considers "factors of unpredictability" connected to the panic and to market trends, citing Max Weber and Jean-Michel Rey. The same concept is remarked by Thrift, who wrote that "these theories are not just about seeking out new knowledge but also about telling stories about an uncertain world which can, however briefly, stabilize that uncertainty, and make it appear certain and centred" (Thrift, 2005, p. 48).

It is clear that the analysis of Bodei is relegated to the modern cultural field, but also, in this case, contains reflections on a global scale and, above all, the idea that the same uncertainty about the future is more visible when considered in relation to international dynamics. There is no lack of references to financial economy, to international political actors and to the inevitable consideration that, in such a historical context, with the loss of every certainty (both in traditional or religious and in ideological contexts), it is impossible to be certain about the future.

In this framework, the idea of modern uncertainty is very clear. If what Bodei states is true, it originated precisely from the decline of traditional and religious societies, which came about, even more vehemently, with the end of the Cold War and of the ideologies of the 20th century which collapsed with the Berlin Wall.

Mercury, fluidity and uncertainty in financial dynamics

The disorder and uncertainty of the world situation after the bipolar system is now considered a reality. We have connected the idea of uncertainty to the rise of what Veca, Minc and Kaplan called "tribalism" and with the concept of "disunity of the world" after the failure of the Hegelian model proposed by Fukuyama. That idea is much more corroborated if we also consider the economic and financial situation in many world regions. According to Fukuyama, economic evolution is strictly connected to the political one: he noticed that "if the process of economic homogenization stops, the process of democratization will face an uncertain future as well" (Fukuyama, 1992, p. 235). It was the similar idea expressed by Benjamin: "the uncertainty of its economic position corresponds to the ambiguity of its political function" (Benjamin, 1999, p. 21).

When considering world uncertainty, it is impossible not to include financial flows, which greatly contribute to current uncertainty. There is a vast literature on the connections between money, finance and uncertainty,[16] which started with Knight (1921), but I will just consider the economic geographical perspective, in order not to go off track. The same uncertainty has been greatly implemented in the recent financial crisis.

Georg Simmel underlined the floating nature of modern economy, emphasizing the difference between the medieval and the modern way of thinking:

> [T]he character of an agrarian economy is determined, on the one hand, by its reliability, by a small and less variable number of intermediate links, by the emphasis on consumption rather than production, and on the other by an attitude focused on the substance of things, and by an aversion to the unpredictable, the unstable and the dynamic.
>
> (Simmel, 2004, p. 235)

One of the main exponents of economic geography who wrote about the fluidity of money is Gordon L. Clark.[17] He gave a strong image of the uncertainty that dominates the world economic context, with direct effects on geographical realities, on different scales: Clark outlined that "money is fluid like mercury" (Clark, 2005).

Even though written before the 2007–2008 world economic crisis, with this conceptual formula, Clark targets some essential elements of the fluidity of economics, which corresponds to the uncertainty that dominates the world scenario. Clark affirms the idea of very cogent economic globalization, which enforces itself day by day. In this context of continuous strengthening, we can distinguish a financial branch and a monetary branch: in this regard, the author underlines, bringing all back within the bounds of the fluidity, that "money is local and finance is global" (Clark, 2005, p. 100), wanting to state the scale shifting from the first to the second economic dimension which occurred during the past century. This kind of linking represents a moment of contact of the flows on a global scale (finance) and the local and national ones. This is a demonstration of an intrinsic overlap of levels when we speak of global finance: not so much a gap, an isolation between them, but a mutual influence. This interaction between the global and local scales helps us to better understand the nature of the fluctuating trends of financial economy and how these can Robert A. Schwartz (Schwartz et al., 2011; Machina & Viscusi, 2014).

Starting from the Clark's affirmation of a direct influence between the global financial element and the local monetary one, and taking for granted Karl Beitel's vision (2000) of a strict correlation between the real estate construction market and the finance one, where the latter is linked in particular to banking dynamics, it is perhaps possible to say that, when global financial dynamics are oscillatory, even the flows connected to them will tend to follow the same trend.

There is no doubt that the metaphor of a currency liquid as mercury well fits into the conceptual formula emphasized here, because it evidences the link between economic and financial trends and the impact on territory which is clearly evident in the cities where these trends are even more crucial than the rest of geographical space, and where the economic "mercury" seems to "coagulate" the most.

In this context of renewed considerations of international relations and of world structures, especially after the economic crisis, it seems possible to connect the idea of "uncertainty" even to the geographical field: as Neil Smith pointed out before this historical phase which started in the 21st century, most if not all of our assumptions about the geographical ordering of the world, from the local to the global scale, are now obsolete, and we find ourselves in a period where theory and political organization have to be reinvented together in order to match new circumstances (Smith, 1999).

On a European scale, the direct effects of the crisis have been evident, from the point of view of job occupation, precisely in the European financial centres: "it was thought that there might be a significant impact on the traditionally prosperous parts of the European economies where many of these jobs are concentrated" (Kitson et al., 2011). Essentially, where finance is concentrated, we can observe the direct signs of the crisis which started in the financial world.

The world economic crisis is a clear symptom of a globalization which connects States from an economic and financial point of view, with a strong connection dominated by fluidity and uncertainty of markets and international stock exchanges.

Uncertainty in migratory flows

If, on the one hand, from the perspectives of political science and sociology the idea of uncertainty seems to be well applied on a global scale, on the other hand, from an economic point of view, some researchers have outlined the connections with the idea of "uncertainty" seen from different situations about international migrations (Allan & Baláž, 2012). This is because migration flows are very often strictly connected to the idea of uncertainty, from many perspectives: the decision on the destination, the effects on destination territory and economy, the moment of return, the job of migrants and the cosmopolitan conformation of cities. Indeed,

> uncertainty in migration has two sources. The first is based on impossibility of obtaining knowledge about current conditions even in the place of origin, let alone in one or more destinations, that you may never have previously visited . . . The second source of uncertainty is the unpredictable nature of the future. All future possible outcomes involve uncertainty, because of the impossibility of knowing the probabilities of particular outcomes in the future, even if they are currently known.
>
> (Ibid., 2012, pp. 5–6)

In Christian Dustmann's writings, the focus pointed out is on migrant savings and the effects on the territory of destination, in a comparison with those of native workers: "savings of migrant workers due to precautionary motives are compared with those of native workers and the impact of uncertainty on migration and remigration decisions is analyzed" (Dustmann, 1997, p. 311). In the author's opinion, the consequence is a situation of substantial uncertainty.

Paul G.J. O'Connell, on his part, highlighted a simple question in his reflections about uncertainty in migration flows: "how uncertainty about these conditions affects the migration decision"? (O'Connell, 1997, p. 331).[18] In his opinion, uncertainty in migrations has a twofold aspect which is inherent to the destination region and the (uncertain) "evolution of conditions" in that territory and in the one of origin:

> [T]he result obtained earlier that increased uncertainty about current conditions abroad spurs migration is more fragile than the result that increased uncertainty about the future deters migration. If the labor market is composed of identical individuals, there is neither a "try your luck" nor a "wait and see" motive.
>
> (Ibid., p. 345)

In practice, standing on his developed model, he put forward two conclusions or predictions:

(a) In a partial equilibrium setting, uncertainty about current conditions abroad can encourage speculative or "try your luck" migration, which is undertaken to learn more about the destination region. This result is familiar from existing search-theoretic models of migration. (b) Uncertainty about the future evolution of both foreign and domestic conditions acts in the opposite direction, leading to "wait and see" behavior. It is optimal to postpone relocation to allow some or all of this uncertainty to be resolved.

(Ibid., p. 346)

With a view to destination decision,[19] Jan Saarela and Dan-Olof Rooth have elaborated an empirical study demonstrating how uncertainty in the first decision-making stages greatly impacts on international return flow. In order to demonstrate this, "high-quality and detailed micro-data" was used in some cases. They concluded:

[U]ncertainty in the initial migration decision might be an important driving mechanism behind the decision to return migrate in the short run. Migrants with a worse-than-expected outcome in the host country upon arrival and shortly thereafter have a notably higher probability of return migrating than other migrants.

(Saarela & Rooth, 2012, p. 1896)

Linking uncertain market flows to migration decisions which contribute to the delay of the same decision, other scholars have written that

when markets are uncertain, families can gain by delaying migration due to the option value effect. This has been recognized in the new migration literature. However, uncertainty can also induce a diversification motive if families are risk averse in the manner well known in the portfolio theory of financial economics.

(Anam et al., 2008, p. 248)

The Italian geographers Raffaele Cattedra and Maurizio Memoli have proposed a particular interpretation of a sort of "geography of uncertainty", as an urban geography of chaos (see Mazzei, 2016), mostly deriving from the confluence of portions of population from all over the world,[20] contributing to reshape a new map of settlements in the most important Italian cities. So, in the authors' opinion, the geography of uncertainty has, above all, a precise scale of reference, which is the urban one, and then, it has some main causes – linked to the immigration phenomenon – with determined consequences:

[F]or some decades, Italian cities, even with different traditions and with more or less intensity and density, have been connoted by the presence of immigrant communities and by new social and territorial practices, which have contributed to modify their appearance, their equilibriums and potentialities.

(Cattedra & Memoli, 2013, p. 143)

From the deriving map, "it seems that a geography of uncertainty may emerge, followed by presupposed social *chaos*, bound by opposite forces, made of additions and exclusions, melting pots and isolation, of opposite identities and also of original 'contamination' situations and cultural complexity" (Ibid., p. 144). The effects on the city, from this point of view, seem to be immediate, because "districts and urban typologies (even in medium size and small cities) appear and disappear, they assume new appearances. Persistances are changed, functions are moved and new territorialities and urban practices are entirely reshaped" (Ibid., pp. 144–145).

In the geographers' opinion, this kind of dynamics has a relevant impact even on the most stratified configuration of the urban asset, and so

> the historical districts become "ethnics" (at least partially), with phases of degradation and new patrimonial values; peripheries take on new functions of public space and of places of aggregation for communities of different provenances; in schools new classes are formed, made up of increasingly mixed students from many nationalities, (in some Northern cities we have 100 nationalities and 50 different languages); on means of transport languages and people from different origins become familiar; in the markets the presence of non-Italian sellers becomes natural, while unusual products and alimentary customs become diffused.
>
> (Ibid., p. 145)

In substance, the geography of uncertainty proposed by Cattedra and Memoli originates from migration dynamics in Italian cities, which contribute to rapidly and directly reshaping the face of the same cities.[21] Therefore, uncertainty is all about the social level of urban social conformation, and so is also about territorial planning and about the structure of urban spaces, which change day by day, and have different profiles and different aspects. These changes will also be based on local and national migration policies, in a sort of mixture of necessary rules and unavoidable chaos: indeed, "the rules decided on by societies, and the continual change in daily practices, formally constitute the order of the underlying chaos and create, contextually, a 'new cosmos'". This "new cosmos" is one "of the new '*cosmo-politan*' situations, which demand the recognition of the right to citizenship to all inhabitants and encourage other speculative investigations" (Ibid., p. 161).

In this case, the association of terms is proposed not on economic aspects, as in the preceding articles, but mostly on migrant flows and urban geography. The impossibility of programming income flows, the "chaos" generated by incoming migrant flows, with their related settlements, and the derived uncertainty in the geographical-urban context are all connected.

This is a very interesting and original reading of the uncertainty topic in a contemporary, geographical context, certainly derived from an acquired asset of world configuration, that has taken on both shape and features, where we can find topics linked to migration flows. These appear, however, with direct effects on urban geographies, as one side of a geography of uncertainty that dominates globally and only afterwards has repercussions on the fabric of European cities.

The reflection of the two authors, also elaborated in a longer essay dedicated to the "spaces of the 'new Italy'", still deserves to be considered, also in view of a thematic contribution that takes into account an even wider perspective which is, at the same time, both chronological and spatial.

Economy, politics and migrations seem to currently live a condition of deep uncertainty, and they are three different aspects of a common process which can be called globalization which did not rightly start at the end of the Cold War, even if that was the moment of "great acceleration", but started centuries ago (see Arrighi, 1994; Braudel, 1949; 1979; Wallerstein, 2011).

Uncertainty as a concept: a philosophical overview

Up to this point, we have tried to give an overview of the different declinations of the term *uncertainty*, as seen by sociologists, economists, historians and, of course, geographers.

The current situation of chaos and disorder, of substantial uncertainty, as many authors have defined it, involves philosophical concepts and considerations, which have seen some authors facing, more or less openly, the topic of uncertainty in a diachronic perspective. Among them, I have already quoted Salvatore Veca, who included in his book *On Uncertainty* (1997) not only philosophical aspects but also linguistic and cognitive ones. He also included translation and interpretation and identity and personal spheres, the sense of truth and justice and freedom and political systems. His philosophical meditations are inserted in the post-Cold War context, in the years in which the common perception was that the international political scenario could find a defined asset, centred on the United States which was however, deeply unstable.

On the concept of truth and relativism, also Sebastiano Moruzzi focused his attention on the definition of *vagueness*, intended as the expression of the *sorite paradox*. Another characteristic of the term *vagueness* is that it has no borders: in Moruzzi's words, indeed, "the apparent absence of clear borders is not only the manifestation of vagueness, but appears also in the indetermination of certain cases when we try to apply a vague expression" (Moruzzi, 2012, p. 10). Uncertainty, in effect, means the loss of a centring order – that is particularly visible in international politics – and conceptually also refers to the loss of truth.

Even in its historical affirmation, that is in any case conceptual, uncertainty could, in some way, be associated with the "tragic" condition of modern man, which has been considered by many authors, from Agostino Lombardo to Carl Schmitt. We will later deal with this type of analysis, which deserves careful thinking and a wider geographical consideration, but that type of relationship had already been pointed out philosophically. Peter Wust, who studied the concepts of Uncertainty and Risk at the end of the 1930s, underlined this relation.

The German existentialist philosopher, while considering the nature of mankind and its received influences, stated in a wide-ranging speech that "there is no doubt that the lack of a homeland, peculiar of man in the order of being, that we are now experiencing, is characterized by a tragic dialectic" (Wust, 1985, p. 71). For Wurst, uncertainty affects the human soul and is closely linked to insecurity, to what the German author calls *insecuritas*, because "even the man who does not shy away from acquiring, through the greatest efforts, in some way, a balance between the two essential powers of his nature, can never reach the status of a completely unassailable certainty of himself". The dialectic between matter and spirit, originating from *humana insecuritas* and from the general shape of indefiniteness of the *humanum*, is tragic and to it "corresponds the subjective side of human *uncertainty*" (Ibid., p. 73).

The connection with man's metaphysical yearning appear almost immediate here, and a relationship with the themes related to the "mad flight" of Ulysses proclaimed by Dante and recalled later seems almost natural:

> [I]n the subjective dialectic of existence, the uncertainty of its certainty has the double function of preserving him [the man] from *hybris*, in the midst of his elevation through reason, and also to lift him from his dust of finite creature to the starry heights of an eternal clarity.
>
> (Ibid., p. 76)

Essentially, as will be better seen below, uncertainty is also related to the thirst and hunger for knowledge of the great explorers. This aspect will be observed later in more detail in the chapter focused on the "mad flight" of Ulysses.

The cause of modern uncertainty, for Wust, derives from full rationality, which leads to a knowledge towards "vertiginous" heights, because "natural reason, with the power of its natural light, can actually penetrate into the deepest mysteries of the world. However, in this extreme abysses, he experiences the mystical catastrophe of uncertainty, despite all the radiant certainty of rational evidence" (Ibid., p. 77).

It almost seems to recall the cognitive process of "*ratio*", which in the long run, starting from Cartesian rationalism, has led man to the modern condition: a light of reason which can lead to the deepest abysses of human knowledge and reason, without finding, in those "abysses", as Wust defined them, footholds and references for human existence. Indeed, in his opinion:

> [O]nly in the religious sphere of supernatural certainty of faith do we begin to understand the profound wisdom placed in the fact that man can never reach the ultimate certainty he craves for himself . . . to become a permanently secure being.
>
> (Ibid., p. 78)

Here lies the innermost essence of modern uncertainty and its paradoxes and contradictions, because the only plausible certainty is then to be found, according to the author, in the sphere of the supernatural, in the links between human beings and God. In this condition, uncertainty retrocedes, opening up spaces to the truth, intended as the deep knowledge of things of the world and of a sense of human existence deriving from the possibility of transcending the rational sphere: where "the intellectual anguish of uncertainty reaches its climax, thus obtaining new strength for the soul, so that man is able to ascend higher than he is able to do solely with *ratio*" (Ibid.).

The problem of *ratio* and its use is central for the author in order to understand the human condition of *insecuritas* mentioned in his work: uncertainty, therefore, becomes a realm of human choice because the same *ratio* also implies the freedom to make decisions, which in turn implies – or may involve – risk. There is an

> internal correlation between *uncertainty* as essential destiny of his being of *ratio* and risk as an essential destiny of his form of freedom of the will. On the one hand, the hazy sight of human *ratio*, with the possibility of error, and on the other hand, the half dependence and the half agility of instinct and human will.
> (Ibid., p. 84)

It is not the case to here consider the conclusion proposed by the author to leave this human condition, which is the result of actions throughout the centuries that have led man to the "mystical catastrophe of uncertainty", but it is interesting to highlight on one hand, the visceral nature of uncertainty as a mystical catastrophe and, on the other hand, the paradox that is the background to all his discussion. According to Wust, after man reached the great heights of reason and full scientific knowledge, sealing the possibilities of human *ratio*, he experienced the infinite possibilities of human freedom, outside any metaphysical order, entering the domain of general uncertainty.

In *Naivität und Pietät* (1925), in a more general consideration of the work of the author, there are references of more direct geographical interest, that will lead back to the following considerations. The author explains the meaning of his book:

> I emphasised the striking contrast that I had strongly felt for thirty years between the idyllic serenity of the countryside and the excitement of city life . . . Man, through his intellectual factor (reflection) is projected out of a biological animal equilibrium. But he is not in a stable situation in the spiritual spheres, as if pure vision could eliminate any anxiety. The human being is a halfway point between bios and logos, a nature in exploration, always on the move, "in via", who lives precariously in a tent.
> (in Wust, 1985, p. 11)

The proposal of Wust, on the projection of uncertainty seen in the light of modern philosophy, captures some essential points that will be elaborated in more detail later. Among these, the most relevant for the purposes of this study is the one concerning the essence of modern uncertainty, which intervenes exactly by attempting to reasonably fix the rational cardinal points, but immediately plunges man in deep uncertainty. This derives, according to Wust, from lack of adherence to supernatural reality: a void that must be filled in order to avoid the dramatic moment of the "mystical catastrophe of uncertainty" and to overcome the intrinsic position "between bios and logos, a nature in exploration".

The paradox individuated by Wust, of "freedom of will" and of knowledge that is intimately linked to modern uncertainty, is also the basis of the cartographic paradox of the Modern Age, as we will see.

If it is true that modern cartographic works started to achieve a high level of "certainty of representation", thanks to the technical innovations and to human capabilities and knowledge, at the same time, they will embody the "representative front" of uncertainty of the modern condition, above all geographic, and of that geography of uncertainty established with the start of Modernity.

Notes

1 See, for example, the Marc Augé case of the *Non-Places* (1995).
2 See, for example, the works of Park et al., 1925; Wirth, 1928.
3 For a guide to the literature on the School of Chicago, see Kurtz, 1984; Linberg, 2008.
4 In his opinion, furthermore, "suddenly everything becomes uncertain, including the ways of living together" (Beck, 1992, p. 109).
5 This idea has been expressed even by Benjamin, right in relation to the uncertainty: "in the flâneur, the intelligentsia sets foot in the marketplace-ostensibly to look around, but in truth to find a buyer. In this intermediate stage, in which it still has patrons but is already beginning to familiarize itself with the market, it appears as the *bohème*. To the uncertainty of its economic position corresponds the uncertainty of its political function" (Benjamin, 1999, p. 19).
6 Indeed, in his words: "the present-day uncertainty is a powerful individualizing force. It divides instead of uniting, and since there is no telling who will wake up the next day in what division, the idea of 'common interests' grows ever more nebulous and loses all pragmatic value" (Bauman, 2000, p. 148).
7 See, for example, the strong ideas expressed by Eric J. Hobsbawm, of the *Age of Extremes* (1995). "In the words of Eric Hobsbawm, as 'the citizens of the *fin de siècle* tapped their way through the global fog that surrounded them, into the third millennium, all they knew for certain was that an era of history had ended. They knew very little else'" (Arrighi & Silver, 2001, p. 257).
8 It had its enormous effects even in the labour market, connected with the idea of globalization, as Blossfeld, Mills and Bernandi have studied (2006).
9 Fukuyama argued that "liberal democracy may constitute the 'end point of mankind's ideological evolution" and the "final form of human government", and as such constituted the "end of history'" (Fukuyama, 1992, p. XI). See also Kolakowsky, 1990; Horsman & Marshall, 1994.
10 See Arrighi & Silver, 1999; 2001; Barnes, 1999; Hannay, 2008; Horsman & Marshall, 1994; Jowitt, 1992; Kennedy, 1993; Rosenau, 1990; Thrift, 1989; 1992.
11 He contests the theories, after the end of the Cold War, which considered the possibility of a sort of "end of geography" (see as example O'Brien, 1992; Ó Tuathail, 1997).
12 In Agnew's words: "depending on the nature of the geopolitical order of a particular period, territoriality has been 'unbundled' by all kinds of formal agreements and informal practices, such as common markets, military alliances, monetary and trading regimes, etc." (Agnew, 1994, p. 54).
13 As he specifies in *Introduction*, "I don't know if History is tragic, but I am persuaded that we should treat it as such, to prevent that" (Minc, 1994, p. 1).
14 He calls it a "disharmonic space" (Minc, 1994, p. 37) and that is different from the European past from the 17th to the 20th century.
15 The same concept is shared by Parag Khanna, who sees the actual confusion, disorder and tension of the world as a new Middle Age: "twenty-first-century diplomacy is

coming to resemble that of the Middle Ages: rising powers, multinational corporations, powerful families, humanitarians, religious radicals, universities, and mercenaries are all part of the diplomatic landscape" (Khanna, 2011, p. 3).

16 Here I quote some of the main essays found at this purpose. See Aastveit et al., 2013; Bhagat & Obreja, 2013; Drèze, 1993; Lensink et al., 1999; Lindley, 2007.
17 He wrote right before the financial crisis of 2007–2008, but he gave a very significant image of the financial flows, which permeate every nation, going over the national borders: this helps us to better understand the global nature of the financial crisis and the connected uncertainty (see also Storper, 1997).
18 He refers to some important authors which contributed to these kind of considerations: Todaro, 1969; Harris & Todaro, 1970; Langley, 1974; Hart 1975; Rogerson, 1982; Gordon & Vickerman, 1982; Maier, 1985; Berninghaus & Seifort-Vogt, 1991.
19 See also Wang & Wirjanto, 1997; 2004.
20 At the migration purpose, from an economic point of view and with connections with the idea of "uncertainty", see also Gaumont, 2012; Czaika, 2012; Dustmann, 1997. In Dustmann's opinion: "savings of migrant workers due to precautionary motives are compared with those of native workers and the impact of uncertainty on migration and re-migration decisions is analyzed" (Dustmann, 1997, p. 311).
21 In their opinion, "in the contemporary cities of the "New Italy" emerges a plural society composed by a variable dimension of *actors*: individuals, families, institutions, enterprises, and even groups, associations from different sectors, communities, *lobbies* (from the volunteering to the criminality), of unemployed, of marginal or integrated ones, poor men, of young and old men, etc.", who "confer to their earthly existence, through the sedentary space that they possess and that have an existential and central value for them, and the forms of circular migration acted by them, as claimed by the new approaches of the social geography" (Cattedra & Memoli, 2013, p. 147).

3 Early Modern European political geography and uncertainty

European political geography in Modern Age

It has been said that the 16th-century left a paradigmatic mark on the crisis of European Modernity and on geographical uncertainty, which was at its base. However, it was also a crisis and a conflict of sovereignty that was expressed in a multiplicity of political forms, of administrative authorities and State entities that, although in a European configuration of clear definition of boundaries, produced a disruption in the field of European foreign policy and inter-State relationships of the Old Continent. This "mayhem", revolution or crisis, is definable according to a logic of substantial uncertainty and even of political chaos.

Political geography, in some way, originated in that political and continental context of Modern Age (see Carta & Descendre, 2008), because during that historical moment, the concept of borders began to be fully associated with State entities: a dense network of diplomatic relations among States was established. This meant that European internal political logistics needed to be rethought together with their internal borders (see Descendre, 2022).

Alberto Aubert rightly affirmed, in a broader discussion about the start of Modernity as the first configuration of a structure of balance of powers (Kissinger, 1994), that

> for the functioning of the system of balance, and in particular of its purposes against hegemonies, it was necessary that the specific aims of the States were all considered equally legitimate, at least formally, since in actual fact the hierarchy of power ended up subordinating the principle of "equal dignity" to the real hierarchy between forces.
>
> (Aubert, 2008, p. 41)

This meant, of course, installing resident ambassadors who could convey legitimate State interests, although the establishment of such a mechanism – already experienced by Italy in the Renaissance[1] – at European level led to a series of technical, organizational and institutional difficulties which were only settled over time.

[But] this institutional fluidity of structures and diplomatic duties – perpetuated at least until the age of Louis XIV, when diplomacy became a defined hierarchy, where a delegate followed the ambassador, and a resident followed the delegate – was in fact only one aspect of a deeper and more general indeterminacy of institutional and bureaucratic boundaries of modern territorial States. It was a far cry from full and effective control of executive, legislative, judicial and administrative functions, at least until the 17th century, and from full enforcement of sovereignty over the still solid feudal, urban and ecclesiastical autonomies inherited from the Middle Ages.

(Ibid., p. 43)

Therefore, in the political system that was developing in modern Europe, made up of balances, embassies and legations bearing national interests – as had already happened in the Italian Renaissance – the institutional structure had two characteristics. Firstly, it was still tied to the dynamics of power, typical of the Middle Ages, at least in the relations between emissaries and trustees: and secondly, it struggled to create a structure of its own, different from what had characterized the earlier periods. Furthermore, it was a still highly unstable, complex system of references, dominated by uncertainty and by the lack of a solid structure that could cope with the renewed problems and modifications also caused by the great geographical discoveries.

This type of uncertainty falls *de facto* within the scope of political geography, of inter-State relationships and of the changing geopolitical structure peculiar to that extraordinary and revolutionary historical moment.

The system of embassies, which will be later consolidated but was at the time still defined "fluid" in consideration of the modern European political and institutional situation, was simply the consequence of the liquidity – of the lack of a supporting structure – of the European political system itself, which had not yet found its bearings. The coexistence of different – more or less equal – State entities was the cause of this "uncertain" or "chaotic" situation.

In such a situation, different pressures came to light, sometimes centrifugal and other times centripetal, contributing to the new state of *equilibrium* – but which was also continually unstable (Nexon, 2009; Rabb, 1975) – in modern Europe. "The sovereignty of monarchs was obliged to battle with, to compromise and not infrequently to share with territorial and class autonomies, since it profoundly influenced the continuous deconstructions and reconstructions of European political order from the Middle Ages" (Aubert, 2008, p. 45).

In other words, the territorial structures were becoming – or had already become – autonomous political formations or organizations, in a clear break with the universalizing tendencies of the past, defining political geography dominated by conflict, by potential and real internal political contrasts and by a balance which was the obvious symptom of an inherent and endemic instability (Rabb, 1975).

This was also definable as a European political uncertainty, since each State has had its own military system, legitimized by a precise and distinct political authority, so much so that the stubborn resistance of class and territorial structures to the inclusive and absolute monarchy ended inevitably by interacting with international dynamics and with the wars, revealing how the interests of feudal lords, of urban patricians, of cities, clergy and the many autonomies scattered throughout the territory were often in contrast with or opposed to the geopolitical interests of the developing State entities.

(Aubert, 2008, p. 45)

Aubert states that all this

contributed to complicate international relationships, already made difficult by the extreme mobility and uncertainty of State borders, altered not only by the continuous change of dynasties and related marriage policies, but precisely by the pressures for autonomy or by the explicit rebellions of provinces, cities and feudal lordships, which occurred in the context of kingdoms by no means homogeneous in territorial, economic, ethnic, cultural, religious and political terms.

(Ibid.)

And,

such conflicts often included what we would now call an "international" dimension. Princes, magnates, and even urban leaders sometimes negotiated, conspired, or allied with outside powers. Rulers exploited internal conflicts to advance their power-political interests and make good their territorial claims.

(Nexon, 2009, p. 6)

What we were saying earlier becomes clear in the following statements about the extreme uncertainty of State borders, that reflected the image of uncertainty of European political structures. Nexon, for his part, explains that

most early modern European states were composed of numerous subordinate political communities linked to central authorities through *distinctive contracts* specifying rights and obligations. These subordinate political communities often had their own social organizations, identities, languages, and institutions. Local actors jealously guarded whatever autonomy they enjoyed.

(Ibid.)

That was the uncertainty of a Europe that was looking for a new equilibrium (Rabb, 1975). It was, above all, a geographical indefiniteness, but it was also a political and geopolitical one: this is why Aubert talks about territorial autonomies, the birth of new States and the renewed political entities, separated from the previous ones, which were trying to establish themselves within a framework of systemic change in Europe. They were connected to the idea of mobility and uncertainty of State borders.

This political turmoil, this confusion of political order, so extremely multifaceted, prismatic, variable and undefined, found itself in a very dynamic institutional reality, similarly uncertain and indefinite, because no longer hierarchical but pluralist, with a multiplicity of State institutions. From this multipolarity, which was also diplomatic, only a multifaceted, indefinite, substantially uncertain reality could emerge which was so even from a social, economic, philosophical and religious point of view.

Uncertainty in the thought of Machiavelli

The situation faced by Europe in the 16th-century had already been seen in the Italian territorial context of the 15th-century (see Somaini, 2013) studied by Niccolò Machiavelli, who was cited by Federico Chabod several times in his definition of the idea of Europe and was also recalled by Aubert several times in his dissertation on the European political configuration in the Modern Age (Chabod, 2007).

Laid the historical basis of that *critical* moment of epochal change, profoundly tied to the idea and concept of uncertainty, instability and research of balances within the European political and geopolitical structures (Rabb, 1975), it would be appropriate to investigate how, at different moments of the 16th-century, it was experienced and interpreted by some great figures of the time. In the early 16th-century, Machiavelli, the thinkers of the second Scholasticism, Giovanni Botero a few decades later, Jean Bodin and René Descartes at the end of the century, all seem to represent some of the pillars of modern uncertainty, for different reasons.

All of them, in one way or another, are representatives of an era of radical revolution, which changed the categories of thought and the same European political reality, making it necessary to introduce or re-think – also because of the European crisis discussed earlier – new conceptual models.

With the emergence of new State entities in the European political and geographical context, Modernity posed new essential questions, not only to Machiavelli, and, later, Botero (2017), but also to other theorists of law, international relations and politics: questions which were characterized by a substantial indefiniteness – mainly due to the phase of profound change that Europe itself was experiencing – regarding the authority of the State and its limits, including territorial limits, the power of the vertices of the State (as in Machiavelli's *Prince*), the collective good and the legal framework within which the power of the State had to be exercised (Cox, 2002).

The revolution of Modernity, in other words, lead European man to essential considerations on theoretical and practical, philosophical and religious, judicial and political, on the course of political action in all its different facets, which had to necessarily include the spatial extent of the power to be exercised.

Machiavelli represents de facto an essential, inescapable landmark of the realist school that had been a cornerstone of the history of political theory since the 16th-century up to the modern theory of international relations, and also for this study, on the (political) geography of uncertainty, as the mentioned authors have widely emphasized.

Machiavelli was one of the founding fathers of realism, albeit starting from ideal considerations on a prince who would have to face the reality of the times, in deep systemic change in which the uncertainties of the modern condition of his readers were further intensified, because he aimed, through his treatise, to "propose himself as a therapy" (Pedullà, 2013, p. XIII).²

The political European reality during the 16th-century was strongly connected to the idea of *crisis, instability* and *uncertainty*, followed what happened, in smaller scale, in Italy during the 15th-century, partly described in the works of Machiavelli.

That European *balance of power* which characterized that historical, cultural and political context, determined a category, at least according to some authors, of European logic and of the concept of modern Europe itself: the internal conflict, the *stasis* (στάσις) (see Agamben, 2017, II, 2), which can, even in this case, be included, in the category of uncertainty.

The work of Machiavelli, centred around a context of profound geographical-political uncertainty, was characterized firstly by the uncertainty of his own personal situation. Giulio Ferroni, when speaking about the success of his works and the easy adaptability of his thought to many different contexts of research, noted that

> beyond the outcomes of these utilizations of Machiavelli as a political handbook and *vademecum*, as prompter of unscrupulous and successful practises of power, it is easy to notice how its historical condition and the circumstances under which his work and thought developed radically contradict any exciting image of political success.
>
> (Ferroni, 2003, p. 10)

The whole work of Machiavelli can be read in the light of uncertainty, on the most different interpretative levels, because

> the whole initiative of Niccolò in his activity as secretary of the republic, and all the considerations, the proposals, the hypotheses conveyed in the works written *post res perditas* are certainly not related to a movement of "rise", to the euphoria of a success or an expansion, but to situations of danger, of uncertainty, of defeat.
>
> (Ibid., pp. 10–11)

How can we draw uncertain geographical or geopolitical considerations from Machiavellian's "idealistic" realism? His book on the prince almost never explicitly addresses any geographical or territorial questions: we can only assume the geographical issues mostly as implicit in the Machiavellian discourse, revolving around the characteristics, qualities and skills that the ideal prince must have to maintain and manage power for the good of the community, in the nature of the States themselves, military art and so on. They basically form the essential background for all the political, social, psychological and "anthropological" considerations on the figure of the prince.

To formulate geographical-political remarks on the work, which until now have almost never been addressed by this disciplinary point of view and in consideration of uncertainty,[3] it is necessary to start from two basic observations in order to gradually move forward: the time in which the *De Principatibus* was written and the premises of the idea, the concept, the theory of "realism" just mentioned.

In the field of politics and political thought within which Machiavelli moved, "the very notions of power, control and safety are continually questioned" and "inside the flames of this laceration, Machiavelli's thought is always penetrated by a very strong sense of contradiction" (Ibid., p. 11). Indeed, he was part of a historical moment of profound change and great indeterminateness, especially concerning the political-geographical future of the peninsula: "in the sense of instability pervading *The Prince* it is undoubtedly possible to see a result of the Italian wars, with their sudden changes of front and with their unexpected victories and defeats" (Pedullà, 2013, p. LX).

Instability, therefore, completely pervaded the Italian political and sovereignty theatre, shaking the whole territory, becoming apparent earlier than in the rest of Europe. It seems possible to define it as uncertain, as both Giulio Ferroni and Gabriele Pedullà have rightly highlighted.

It was therefore a historic moment "in which everything is unstable and nothing lasts" (Ibid., p. LX), in which the myth of the goddess Fortune would necessarily play a key role, because of the deep uncertainty that dominated the political landscape: "blindfolded, with a habit of two different colours (to indicate its double face, good and bad), balanced precariously on one or even two globes" (Ibid.). *The Prince* was developed in a few months, published for the first time at a distance of about 20 years from the enterprise of Columbus, which was, it should be remembered, geographical and political at the same time, and which was the symbolical, but also actual start of the Modern Age. The prince lead his action within the decadence of nobility and, with it, of the idea of medieval chivalry – based on certainty – and Machiavelli had to deal with it to delineate his characteristics.[4]

In this context, "the same movement that threatens to make us fall is the condition of political possibility in a world in which none of the bodies is ever at rest (a reminiscence of Lucretius?)" (Ibid.) and this is just because "Machiavelli described a state of widespread instability, where the reference to *necessitas* risked becoming so frequent as to cast doubt on the value of the same law" (Ibid., p. LXIV).

The fragmentary dispute, therefore the fragmentation: here is one of the key terms that would later find a European seal with the Peace of Augsburg in 1555 and with the affirmation of the principle of *Cuius Regio, Eius Religio*. The subdivision had already emerged in the years of Machiavelli's "deep disconnection of noble domains" (Chabod, 1964, p. 48) which would lead to the "*Foederatio Italica*" (Ibid., p. 15), and the consequent attempt of coping with the subdivision and fragmentation of the territory, symbol of the latent uncertainty in the political and geographical Italian landscape.

Princes and kings were and had become the exponents of that situation, which would be further discussed by Machiavelli in the *Discourses on Livy* (2021), in his exaltation of the idea of the republican form. In that "world of academics and indifference [that was the world of Machiavelli] – underlines Chabod – the prince was the only lively figure, and diplomacy was the only open field" (1964, p. 41) of action for the prince himself. Niccolò "does not have to search very far" (Ibid.) – continues the Italian historian in the famous introduction to the edition of 1961 – to find the primary characteristics of the ideal figure of the prince who stands out at that moment, firstly in Italy and afterwards in Europe: "in that crowd of little lords and adventurers saddening the cities of central Italy, he [indeed] finds the scattered fragments of his prince, the individual parts suitable to be recreated in a more complete and more consistent figure" (Ibid.). These "little lords and adventurers" become representatives of a political and territorial diversity that would be expressed, finding its natural structure, in the idea of *balance of power*, which would become a hallmark of the modern state of uncertainty (Paul et al., 2004).

Italy was the first geopolitical theatre, the early test-bed of what then happened in the rest of Europe a century later, in the propulsive explosion of Modernity of the 16th-century, and that would bring Carl Schmitt to say this about *Jus Publicum Europaeum* (2003).

Machiavelli, not by chance, right in the first three lines of his treatise, claimed to see a world in which "all States and all dominions that have had and continue to have power over men have been, and still are, either republics or principalities. Principalities are either hereditary, in which instance the family of the prince has ruled for generations, or they are new" (Machiavelli, 2005, p. 7). All States, which govern the lives of men, are republics or principalities, where lords and princes govern.

Already in this first statement, we find a fundamental characteristic of that time: the idea of the State, which would then become the nation. It had political power which identified itself with a defined territory and where men form a community lived and made up a people. A people can be defined as such when it has an identity, which is confined within a given territorial area, where the State, through its actors, exercises its legitimate coercive power.

In this context of political changes, the national interest replaced the medieval concept of universal ethical principles,[5] becoming "the guiding principle of European diplomacy" (Chabod, 1964, p. 41): here originates the Machiavellian discourse about the ends that justify the means, for the common good, giving centrality to the State, in a radical transformation of the political logic underlying the action of the prince and of Italian, first, and then European sovereignty. And this brings with it the concepts of territoriality, of community, united under the same banner, to guarantee – quoting Thomas Hobbes – individual securities, the only security understood and conceived at the time, at political levels: the State security. A security, however, that was subject to the continuous questioning of its sovereignty, precisely on the basis of the principle of balance that is synonymous of a state of perpetual instability because

> instability is the condition of all the recent powers, and the *Prince* aims to help Lorenzo to overcome this first, still fluid, moment by rooting his "state" in firm ground, but above all, by teaching him to recognize the water from the mainland.
>
> (Pedullà, 2013, p. LXXVI)

Machiavelli's problem is therefore, also, to propose – even through the idea of strength – a *solution* to the fluidity, uncertainty and indeterminateness of the Italian political condition of the 15th-century, a firm and stable solution within the framework of the more general uncertainty besetting the same Italy. The prince must act in that context, and for this reason, he would have to use all his best qualities, to deal with that situation of indeterminateness and instability:

> [I]n a world dominated by the inconstancy of phenomena and appetites, strictly connected to the problem "laying down roots" of "stopping", in all its different meanings: stopping "a government" (*Principe* VIII), stopping "the attitudes of its citizens" (X), stopping opponents, as did the impetuous actions of Giulio II (XI) or generally stopping his "state".
>
> (Ibid., p. LXXVI)

Firmness and making things stable for Machiavelli meant, above all, a guarantee of certainty, which was the most important thing to establish in the military field: "the principal foundations of all States, the new as well as the old or the mixed, are good laws and good armies" (Machiavelli, 2005, p. 42) and, at the contrary, "if a prince holds on to his State by means of mercenary armies, he will never be stable or secure" (Ibid.). Therefore, "the things written above make a new prince seem like a long-established one, and render him immediately more secure and settled in his State than if he had possessed it for a long time" (Ibid., p. 82). This certainty also passed through the autonomy of action, which was also well-analysed by Pedullà.

The defence of the *raison d'État*, which would become of national interest over time, is first of all a defence of the boundaries within which the prince acts (Meinecke, 2017). Here, the question of the birth and rise of diplomatic missions arises, because these are borders conquered and torn down by war, peace, agreements between parties (Machiavelli states, in Book III, that "the invader has no trouble whatsoever in winning them over, since all of them will immediately and willingly become part of the state he has acquired" (Machiavelli, 2005, p. 12), in a sort of international "anarchical society" (Bull, 2002).

The society described by Machiavelli in his work had originated as a result of the *finis imperii*, the decadence of the imperial principle, herald of certainty and organizer of everything in the European Middle Ages: the multiplicity of Nation States, firstly Italian and then European, opposed to unification and certain principle of centralization of imperial conception. As stated by Henry Kissinger in *Diplomacy*:

> "with the concept of unity collapsing, the emerging states of Europe needed some principle to justify their heresy and to regulate their relations. They found it in the concepts of *raison d'État* and the balance of power. Each depended on the other. *Raison d'État* asserted that the well-being of the state justified whatever means were employed to further it; the national interest supplanted the medieval notion of a universal morality. The balance of power replaced the nostalgia for universal monarchy with the consolation that each

state, in pursuing its own selfish interests, would somehow contribute to the safety and progress of all the others" in the continuously fluid and changing political context.

(Kissinger, 1994, p. 58)

The era of Machiavelli saw the rise of the political phenomenon of the relationship between State entities, which will characterize Europe of the 17th century and which had already started in Italy a century earlier. The history of Italy in the 14th and 15th centuries, also used as reference by Machiavelli, is indeed the history of wars between cities wanting to expand centrifugally from the "dominants", coming to inevitable conflict, as each of these centres had to clash with the will of the others, in a paradigmatic Clausewitz-like logic, everyone impacting with the force of expansion against others. That's why they had to find a way of coexisting, which would be expressed in the concept of political *balance of power* which clearly emerges in the reflections of *The Prince*, imagined as the one who brings together those broken pieces and "essentially restores order" (Chabod, 1964, p. 49). This missing geopolitical order attests the rise of uncertainty in the same field, because when the established order is gone, disorder takes hold, with chaos and uncertainty.

"Equality – in this game of relationships – was in the hands of diplomats" (Ibid., p. 54), who would become, as has already been pointed out, the real actors in the scenario of European international relations, also to maintain the balance of which we are concerned. All this happened in the world studied and observed by Machiavelli, and would occur later, tragically, in the rest of Europe.[6]

This is, in fact, the tableau of the Italian political scene, as in *The Prince* we have an outline of the political and relational theatre of Italy at the time, which is faceted, varied, ever-changing and fluid:

> but it is in the new principality that difficulties arise. In the first place, if it is not completely new but is like an added appendage (so that the two parts together may be called mixed), its difficulties derive first from one natural problem inherent in all new principalities.
>
> (Machiavelli, 2005, p. 8)

The discourse here is very clear and refers precisely to that state of profound instability which is characteristic of Italy at the end of the 15th-century.

It is a theatre which would then be transposed in the European cartographic field with the concept of *theatrum mundi*, which is the title of the first atlas of the late 16th-century (Goffart, 2003). Theatrical performances in which man was perceived as an actor, an integral part of a collective staging, in a context – paraphrasing William Shakespeare – where "the world is a stage, man is an actor and where everyone plays their part". According to Machiavelli, in a world that was finding new and renewed forms of political action and thinking, in an ever-changing relationship between States and internal hierarchies to different contexts, the prince had to play precisely the role of the sovereign. Now, without the perception of unlimited political space of imperial conception – which would remain in the formulation of

Monarchia universalis of Charles V (Helliott, 1964; Pagden, 1995) and Philip II (Parker, 2001) – namely of a set of borders to be crossed to affirm that universal principle, the political perception of territories changed profoundly, including the modern political and geographical sense of uncertainty.

The empire is by definition limitless in space and time, and therefore, any geographical demarcation limits its action. The territory and all its boundaries are therefore seen in a negative light; by contrast, in the modern conception of the State, the demarcation and clarity of territorial borders are considered positively, as elements contributing to the pursuit of power because they guarantee the action of State law, establishing the geographical lines within which the power may be exercised and legitimate coercion be used. They also supply a geographical scenario to the community that lives there and that must recognize itself within those territories. The boundaries, the identification with them and the defence of territory became the interpretative keys used by Machiavelli before others in his political analysis. This is the undercurrent, perhaps unconscious, of his analysis in *The Prince*.

In this new, modern context, the State and the sovereign were absolute rulers in accordance with *legibus solutus*, free from the laws that normally regulated social life. No longer absolute as in the medieval world[7] which aimed to fulfil a divine destiny, wanted and ordained by God, but to honour the only task that the modern world, with its secularizing thrust, required: the management of the State, of temporal things, for the freedom of individual members of the social unit. Princes were namely *principes libertatis*, to use the words of Machiavelli.

These comparisons between emerging political entities, which at the beginning of the 16th-century were mostly typical of the Italian context, would become, at the end of the century, the *Universal Relations* described by Giovanni Botero (Botero, 2015; see Descendre, 2022; Raviola, 2015; 2020). Universal, because the dynamics of the Italian world were reintroduced on a global scale, in a context which would cause the birth of "global linear thinking" theorized by Schmitt (2003).

The analysis of the Jesuit Botero had been defined by Claude Raffestin as "a milestone in the development of political geography and also of international relations" (Raffestin, 2012, p. 55) because Botero studied the world from geographical and geomorphological data and adapted it to the political actors, thus maintaining the prince as a central figure.

However, if Botero at the end of the 16th-century was one of the initiators – if not the initiator – of modern political geography, we could anticipate these thoughts by 70 years and apply this same label – as we are in fact concerned with a label – to Machiavelli, because that *raison d'État* cited by Botero, centred on territorial (and uncertain) data, was the fulcrum of Machiavellian considerations in *The Prince*.[8] According to the observations of Federico Sanguineti (1981), the desperate attempt to have a strong hold on every aspect of popular and national life was the same attempt which Antonio Gramsci made in prison, and it is realistic and confined to idealism at the same time. They started from concrete facts, the "irrevocable reality", as Schmitt defines it in *Hamlet or Hecuba* (2006, p. 39), that would lead other authors to consider Modernity as a rebirth of tragedy, as an extreme proposition of reality, without ideals and without religious or metaphysical escape.

Based on that "irrevocable reality", and on the same realism, Machiavelli outlines the features of a prince based on that factual reality who must know how to manage power and to govern his dominions for the collective good, in a context of extreme political and geographic uncertainty.

The territory and its cartographic transposition represent, therefore, an axis around which necessarily the political dynamics of power and supremacy in the Italian 15th and early 16th centuries lived by Machiavelli revolve, and in a Europe of that same 16th-century, both dominated by uncertainty and by attempts to establish a political equilibrium, each sovereign and prince had jurisdiction and were legitimized in using force on their territories. Each sovereign was party and judge in open conflicts, which were conflicts for the defence of their own territories and also for the acquisition of new ones.

At this point, territories had to take on a central role in political thinking. This was the moment where the analysis of *The Prince*, and of the treatises which at the time considered the rise of new sovereignties central, became geographic and political. The distinction with the imperial conception was then the fundamental key to understanding, in the Machiavellian Modernity, that space and its boundaries were used primarily to distinguish the different areas of State authority. This change was well-explained – better than anything else – by the modern readaptation of the medieval formula *rex in regno suo est imperator*: State territoriality was therefore, reborn (Elden, 2013), the actual relations between States were returned and, with them, so was the distinction between domestic politics and foreign policy. We could already talk about a System of States and of relationships between State subjects, each of which had to necessarily identify with their territories.

Territory was always present, geography existed and fully reappears – it could not be otherwise – in the realistic considerations that take into account international "anarchy" and chaos caused by the simultaneous presence of multiple State actors.

It is implicit in Machiavelli, while Botero is more explicit (Raffestin, 2012). It was certainly a long time from their analysis to the birth of scientific disciplines such as political geography (about three centuries passed), but that time is much shorter than we might imagine, not only in shaping political geography but also in shaping political geography of uncertainty.[9]

International political issues in the rise of the Modern Age

Almost contemporary to Machiavelli, the theorists of law and State included in the so-called Second Scholasticism were able to bring up pressing issues in the very early part of the 16th-century. They mainly focused on the European race towards the New World, while maintaining their vision rooted in European political forms in the Old Continent. They asked themselves:

> [I]s there a right to revolution? Is the violent suppression of a tyrant as a result of a public verdict legitimate? Is private initiative lawful? Should a

door be left open to prevent political power from degenerating into tyranny, or should it be definitely closed to avoid insecurity and instability of power and civil society?

(Giacon, 1950, p. 7)

These issues were clearly of primary importance in the European context at the beginning of the 16th-century, in terms of continental domestic policy, and not only, which was characterized, as reaffirmed by Carlo Giacon, by an absolute social and political indefiniteness, such as to induce thinkers of the time to attempt giving clear boundaries to the political and legal doctrines that were taking place on a global scale.

It was this very same opening up of Europe to give rise to legal considerations which were not bound by defined limits but which tried to fit in with international world politics.[10]

The problem was in fact crucial for Europe, not only when it came out of the medieval and then feudal world but also in the aftermath of a world war (Bloch, 1961; Sassen, 2006), fought on home ground.

The focal point of philosophers, political scientists and legal theorists at the beginning of the 16th-century was directed towards the very urgent contingency of a European continent strongly fractionated internally, where national sovereignties more or less geographically defined but certainly contributing to the state of instability and continuous fragmentation, asserted themselves.[11]

In this context of new dimensions, of geographical discoveries and of political and religious upheavals in Europe arose considerations on the State, on sovereigns and their rights and on international law. This occurred mainly because

the medieval concept of the Holy Roman Empire which should have united the Christian population politically into a single state, already united in the ecclesiastical sense, had evaporated due to frictions between emperors and popes, and the autonomy of national kings from the German emperors.

(Giacon, 1950, p. 29)

Beside these political and religious problems within Europe "with the discovery of America and of the new ways of communication with the East, other particular problems of international law arose" (Ibid.).

Francisco de Vitoria was among the major theorists belonging to the Second Scholasticism, who faced the great questions posed by the discovery of America and not only, like the legitimate authority of the *potestates*. He called into question the political premises deriving directly from the discovery of the new continent, which were still the result of a universalistic medieval conception.

The ideas and actions of Scholasticism, in fact, opened the way to different interpretations. They tried to systematize and objectify the law after the events of the Reformation. They represented "not only the Catholic response to the Reformation, in terms of objectivity as opposed to subjectivity, of rationalism opposed to voluntarism, but were . . . also the vehicle of a modernization and rationalization of European thought" (Galli, 2005, p. VIII).

In the pursuit of that objectivity, however, together with the attempt to give a definite structure to that moment of deep systemic change through the return to Thomas Aquinas and of the theory of *lex naturalis*, as well as of Aristotle, political frameworks which were trying to politically systematize the openings towards the New World were deeply undermined: "according to Vitoria, the pope is not *dominus orbis*, and . . . therefore he cannot legitimize the political domination of a prince on old and new lands" (Ibid., p. IX), because human communities are naturally perfect. That argument would be opposed by Charles V, who nevertheless had deep personal esteem for the theologian of Salamanca.

In this context of debate on the concept of *justum bellum*, just as for Machiavelli and Bodin in the European political scene, territory and its sovereignty was essential, even in view of the uncertainty mentioned several times above. In the case of the new discoveries, according to Vitoria,

> neither the title of the discovery nor the lack or refusal of faith on the part of the Indians, nor their condition of sin or their scarcity of reason can justify Spanish domination or deprive the Indians of their status of legitimate lords and masters of their territory.
>
> (Ibid., p. XII)

The wars against the Indians were considered justified even if they had trampled on universal rights to preach the Gospel and evangelize or had freely transited and traded on their territories, as they were in the case of tyranny based on the killing of innocent lives and human sacrifices.

Basically, although Vitoria was interested in providing general criteria, he avoided dealing with the specific issues of Spanish rights of possession and authority on American lands: "his position is that the Spanish domination in America is legitimate, although not for the reasons that are usually believed, while many of the forms of that domination are probably illegitimate" (Ibid., p. XIV).

The concept of a just war was therefore deeply undermined, no longer justified by a *jus* of divine origin but by the *lex naturalis* and by the fact that the Indians were an obstacle to the evangelizing action of the Spaniards.

The same political framework of thought of the European political system would experience a moment of radical rethinking and change: if for Machiavelli and Botero, the change regarded the internal structures of the Italian or European context, referring to the mode of action of the new real political actor, the prince, for the thinkers of the Second Scholasticism the problems concerned the changes in European political logic, especially in reference to the extension towards the New World and to the possibilities of projection of European sovereignty in those lands. In this latter consideration, strong criticism of the Spanish imperial system emerged, which contributed to an underlying uncertainty which included the geopolitical thinking of the time.

54 *Early Modern European political geography and uncertainty*

We can also find this change of thought and political categories in Bodin (1579), who could be placed as an "intermediate" thinker between the certainties of the Middle Ages and the uncertainties of the Modern Age, leading directly to a deeper thought – which would be better solved by Descartes – on the borders and their uncertainty in the Modern Age.

Bodin: from medieval certainty to the uncertainty of the Modern Age

Bringing knowledge, which had sprung from divine dimensions, onto a human plane, together with a spatial revolution, brought about by the discovery of America and the political changes in Europe, had repercussions which substantially changed the cultural attitudes and ideas of European man.

The revolution of Modernity also stringently concerned the geopolitical field, as we have seen, because central power of imperial authority eroded over time, even if it was still present in certain forms and was definitely in a critical position (Aston, 1965; Kamen, 1971; Parker & Smith, 1978; Trevor-Roper, 1967). This was the case of the *monarchia universalis* of Charles V (Helliott, 1964; Pagden, 1995), who had to struggle with very strong centrifugal forces, as in the case of the United Provinces, configuring an additional battlefield in the political and religious European arena (see Schama, 1988).

However, in a comparison of imperial universalisms such as, for example, that of Charlemagne and subsequent ones, we can glimpse very significant differences in the Modernity of Bodin. They are observable before any other element in the geographical field, namely on territorial claims, on territorial possessions and on the possibilities of use of power within a larger or smaller portion of the globe. Attaining to the specific example cited by Margherita Isnardi Parente, "Charlemagne used real and effective power and territorial possession, a territorial possession so vast as to embrace the widest part of the known world" (Isnardi Parente, 1964, p. 59), which did not occur in modern empires – or, at least, not to the same perceptive extent.

The imperialistic tendency of some of the political forms of the modern world lived in by Bodin were both limited territorially and in the expression of their authority (Tilly, 1994). Above all, and understandably, the area of sovereignty was limited due to the emergence of a multiplicity of expressions of sovereignty during the 16th-century, starting from the Reformation process that contributed to the transformation of the political European map (Trevor-Roper, 1967).

The "imperialistic" logics, in other words, not always disappeared, but they simply had to adapt to the profound geographical and political changes occurring in the Modern Age (Pagden, 1995), and thereby limit their action (Colombo, 2014, p. 53), as they were confronted with equal entities or at least not inferior, in accordance with imperial State logic (Bull, 2002). Insardi Parente highlights this different approach to territorial sovereignty in her introduction to Bodin's *Les Six livres de la République*.[12]

That is why uncertainty is not found in simple objective fact in the presence or absence of boundaries determining a limitation in State sovereignty in the Modern Age: its underlying logic is more important, and the emphasis should be put on that "lack of power" cited by historians, which is actually the core of political configuration of modern States, in some ways, radically distanced from imperial logic.[13]

Therefore, the multiplicity of the political European framework (Tilly, 1994), determined above all by the Reformation and by the decay of the imperial model, would then lead the States to delineate their line of action within boundaries as defined as possible (Krasner, 1999). It is also true, however, that this quest for certainty was caused precisely by decaying power, by the absence of territorial control, which had been previously stable and included in imperial canons.

These statements, and also the aforementioned one made by Aubert (according to which the modern structure of borders was uncertain and mobile), are debatable if we cite, for example, the mobility of imperial borders and the lack of spatial definiteness, precisely because they tended to universality overcome any delimitative attempt, both spatial and temporal (Gregory, 2008; Schiavone, 2008). This topic will be dealt with later, at the end of this chapter.

Considering the passage from the transcendence of an imperial vision to the immanence of the modern conception, "this superiority is no longer, for Bodin of course, the sacred universalism of the medieval empire. It is a real superiority, a power used upon a real and effective territorial possession" (Isnardi Parente, 1964, p. 59).

Also, at the base of his political logic, Bodin considers a series of natural elements no longer as imperial logic would have them, a transcendental vision oriented towards the supernatural. This is the basic difference between the transcendental vision of medieval times and the concrete one of Modern Age, as Aron J. Gurevich points out: "medieval master craftsmen, writers and artists ignored the visible outlines of the earthly world surrounding them and kept their gaze firmly fixed on the world beyond. The result is a highly distinctive, way of seeing things" (Gurevich, 1985, p. 5).

That is why this transition from transcendence to immanence (see Simmel, 2004, p. 78), which is how it has been rightly defined, implies an intrinsic "uncertainty", precisely related – in the thought of Bodin – to natural dynamics:

> Bodin includes the events of the States and constitutions in a general context of subdivision in climatic zones and, consequently, of natural differences of human types and characters, within a broader structure of zodiacal variations, dividing the land in areas corresponding to the various celestial signs.
> (Isnardi Parente, 1964, p. 70)

It is clear, therefore, that such a structure is no longer "transcendentally" stable, reliable and clear (Gurevich, 1985), but must be related to geographical position, to the dynamics of the Earth and its different configurations, almost in an early attempt of geographical determinism, which can fit into a broader discussion on the "geography of uncertainty", especially considering that the "optimum" does not derive from "metaphysically" sure factors, but, "compared to human possibilities, it is achievable by creating a harmonious government, therefore translating it into geographical terms" (Isnardi Parente, 1964, p. 72).

In other words, Bodin considers the optimum justifying everything in somewhat arbitrary terms, because related to the terrestrial latitudes and longitudes, which are impossible to consider in imperial logic, although the elements of his mindset are "all supported by a continuous tradition of thought that has its roots in the Middle Ages, indicative of a philosophical thought that inserts recent theoretical and political reasons on the body of a tradition still felt and experienced with full participation" (Ibid., p. 82).[14]

The role of nature in the government of States and internal relationships is once again emphasized by the author, who remarks that

"a monarchy, an aristocracy, or a legitimate democracy is the one that fully complies with the laws of nature, namely personal freedom and private property of its subjects, assets whose sacredness is precisely sanctioned by nature itself, which governs by means of established laws", because "in relation to its compliance or non-compliance of the laws of nature, a regime manifests itself as legitimate, tyrannical or despotic".

(Ibid., p. 89)

Here nature is clearly intended as the state established by the individual and therefore collective freedom, but that seems always subject to variable social and temporal conditions, although Bodin is a thinker at the turn between the medieval and the more markedly modern logics: his thought is actually

a mixture of medieval tradition, still deeply felt, of learned and mystical humanism, astrological and cabalistic culture, of juridical awareness of the situation and on practical compromise, political sense of power seen as strength and deceitfulness and religious-metaphysical sense of power as an act of freedom which governs the universe, of unscrupulousness and reverence, of universalism and passionate prejudice.

(Ibid., p. 99)[15]

Descartes, going beyond Botero, "would be the expression of the new cultural epoch born in France and in Europe in the 17th century, an expression that is here legitimately opposed, with some schematic forcing, to the thought of Bodin, almost as a symbol" (Ibid., p. 99–100).

The substantial passing of an era, from Bodin to Descartes, took place in this historical, philosophical change of mindset, of *forma mentis*, as symbolized by these two huge characters. It produced true modern uncertainty (from which they tried to escape). Its dawn had already been sen in the works of Bodin where the modern cartesian philosophy was clearly implied.

In Bodin, uncertainty can be found in the discourse on nature that appears less solid than in most previous thinkers, in those natural and geographical dimensions considered *decisive* factors in political relations between States and their governmental configuration; in Descartes, however, as will be seen later, uncertainty seems to form the conceptual basis of his thinking, with direct impact on the political geography of Europe.

We must now deal with Descartes, albeit briefly and rather succinctly, and his mindset, which formed the basis of modern political conceptions and of modern life itself, and which, as Wust suggests, well describes also the current human condition. After a brief analysis of his thought, I will try to apply one of his metaphors to the ideas onto the geography of the current liquid boundaries, and see if, in the long run, it can be linked to the dynamics of the Modern Age.

Descartes and the (philosophical) sublimation of uncertainty in Modernity

With Machiavelli and Bodin, Modern Age politics was already, in short, deeply rooted to certain ideas of indefiniteness caused by critical and systematic changes in the Modern Era, which could also be seen in Descartes' philosophical, modern ideas of full Modernity, where we can also find a sense of sublimation of modern uncertainty.

The same assumption on which modern philosophy is partly based, the Cartesian *cogito ergo sum*, is, by definition, the declaration of the state of uncertainty of modern man. It seems to have a common thread with what has been reported earlier – namely the process of immanence, of movement from a transcendental approach towards the reduction to human intellect – and Cartesian logic. A thread that is the harbinger of substantial uncertainty in philosophical thought, and yet appears to also permeate political and geographical fields.

Steven Nadler wrote about Descartes that "the philosophical authors he was forced to read in school seemed to have abandoned the search for certain knowledge and showed more deference to authority – both to Aristotle and to the Church – than to truth" (Nadler, 2013, p. 14).[16]

At the same time, it is also true that one was faced with what could have seemed another huge dilemma, at least, in the philosophical field, because in Descartes' work, there is the search for "certain and evident cognitions". With regard to the *Regulae ad directionem ingenii*, Nadler states that "the work is, in effect, a manual for how to achieve true and certain knowledge in any discipline". For this purpose, in fact, "mathematics, primarily arithmetic and geometry, represent the paradigm of knowledge because of its method of starting with absolutely certain first principles and proceeding by lucid demonstrations leading to indubitable conclusions" (Ibid., p. 24).

Now, the question is: how can a method that aims to reach definite conclusions, which laid the foundations of modern scientific methodology, on a path leading to indubitable conclusions, be considered one of the pillars of modern uncertainty, which is also then reflected in the geographical field?

Despite being one of the founding fathers of the modern scientific method, Descartes represented the deep crisis of the system of certainties of the past, as he was a symbol and contributed to the process of immanence mentioned above, which overturned transcendental ideas of medieval logic and imperial politics: "Descartes is confident that other sciences – all of 'human wisdom' – can acquire that same degree of certainty, well beyond the mere probabilities (and, hence, controversies) provided by Scholastic thinkers" (Ibid.).

Actually, it would seem a huge paradox to encompass the theorists of the Second Scholastic and Descartes in the same uncertainties, even in the light of what has been reported by Nadler. But going beyond mere assertions, it is necessary to consider that the same continuous search for certainty, even of the Cartesian method, is actually an attempt to provide for the fundamental *vacuum* of certainties which occurred after the end of the Middle Ages.

All the work of Descartes is focused on the search for truth, no more seizing on metaphysical entities but relying on his own reason, so much so that *his personal* point of view clearly emerges. The author highlights this repeatedly, for example, when, in the first pages of the *Discourse on the Method*, he states: "my intention has never been to do more than try to reform my own thoughts and to build on a foundation which is entirely my own" (Descartes, 2020, p. 36).

Here, already, we have an appearance of a clear change in the perspective that had already been established in the minds of modern European man, in a vision that was no longer universalistic and tending to universalism, but was based on individual reasoning and rationality, exclusively tied to the individual. Mentioning his early experience and his studies, still in the *Discourse*, Descartes claimed: "I found mathematics especially delightful because of the certainty and clarity of its reasoning. But I was not yet aware of its true use" (Ibid., p. 31).

No statement could be more telling. Modern philosophy is a continuous search for certainties lost over time. The world and the system of certainties on which the whole cultural system of medieval religious beliefs and divine references was based is gone. In the uncertainty that the modern revolution inaugurated, precisely from the journey into the unknown, the "mad flight" of Columbus, there is necessarily a continuous search for certainty.

Later in the *Discourse*, Descartes says that it was necessary to undermine the system of truth of the past to reach the truth of reason, in an attempt to establish new certainties on human reason:

> I convinced myself that the best possible thing for me to do was to undertake to remove them once and for all, so that afterwards I could replace them either by other, better ones or perhaps by the same ones, once I had adjusted them to a reasonable standard.
>
> (Ibid., pp. 35–36)

The idea of Descartes is therefore to undermine the very concept of dogmatic acceptance of religious truth taken as absolute. As he states,

I firmly believed that by this means I would be successful in conducting my life much better than if I built only on old foundations and relied only on principles which I had been persuaded to accept in my youth, without ever having examined whether they were true.

(Ibid., p. 36)

In order, then, to reach the truest possible conclusions, once challenged the past belief system (accepted as true without any doubt or logic filter), he would have followed four logic cornerstones, the first of which was: "I would not accept anything as true which I did not clearly know to be true" (Ibid., p. 38). Reason necessarily had to filter any kind of answer or statement which had not been verified before "clearly and distinctly", passing through the "cogito", and therefore through doubt.

Modern sciences and philosophy, which found in Descartes a central stronghold, are based on this assumption: they unseat the belief system of the past, do not ever accept them as such, and found a new method precisely based on doubt, on the undermining of any kind of religious "certainty" of the old medieval world, founding new certainties on reason, on doubt, on rationality of course understood in an individual sense.

All this is rooted in a key concept of human existence, of the thought of man and his intellectual faculties: *truth*. If the only truth accepted before was the one of God, Descartes and modern philosophy inaugurates the era of individual truths, the basis of individualism, derived from a critical examination of individual reason.[17] But precisely the doubt, on which the whole philosophy of Descartes and of Modernity is based, bears uncertainty in itself, even in the search for certainties. And indeed,

> Descartes makes a first choice of criterion: to destroy his opinions, he does not think of having to refute them one by one, but he will see as a sure sign that an opinion should be rejected if there is any possibility of finding "some cause of doubt" [in it].

Thus, "in other words, doubt becomes criterion of falsity *a priori*" and among the problems that Descartes does not address is "that to affirm doubt as a sure sign of falsity is an option, not philosophically discussed, that excludes many others for which at this level, it would be possible to opt". For example, "doubt could be seen as a sign of uncertainty of the knower, or as a sign of the need for further investigation, and so forth" (Urbani Ulivi, 2001, p. 25). And again,

> [T]he path of intellectual discovery of the *Meditations* does not allow the use of definitions here, such as those of the *Regulae*, but puts the reader in a position which permits him to directly and personally seize a new ontology, made up of simple and universal things, which remain behind the changing and transient world of the senses.

(Ibid., p. 28)

Here, it is possible to discover one of the sides of modern uncertainty, that right on the "changing and transient" world of perceivable things have its certainties based, so much so that "Descartes eradicates the certainty of mathematical truths" assuming that "all human knowledge, even of maths, could be mistaken" (Ibid., p. 29).[18] Such doubts on world certainties also involve those who believe in God, for whom "the founding of epistemological beliefs remains as it is compatible with a belief in God. Completely devastating human knowledge" and as such "a doubtful challenge cannot be proposed within an atheist context but must be proposed within a religious one" so "the certainty of knowledge remains (. . .) even from an atheist point of view" (Ibid.).

In effect, "from both a believer's and from an atheist's perspective, the epistemological assumption that all human knowledge is erroneous remains possible" and "Descartes therefore pushes the mental experiment to the point of considering doubtful things no longer simply as doubtful, but even as false" (Ibid., p. 30). Still on the certainty and uncertainty of doubt,

"Descartes has not assigned any limit to certainty or doubt: doubt, for the indeterminacy in which its meaning was left, has no precise identifiable boundaries, and also as regards certainty, there is no indication of the extent of its claims", and "if doubt has endless possibilities of practise, and certainty is claimed as infinite, the sceptical conclusion is implicit in the premise, in the statute of "doubt" and "certainty".

(Ibid., p. 32)

Essentially, even more clearly, "looking for an anchor of certainty in the 'deep vortex' of universalized doubt", Descartes seems to find it "in the affirmation that 'nothing is certain'" (Ibid., p. 33). The search for certainty in the boundless uncertain world of Modernity is clear in the words of Urbani Ulivi, when she says that doubt "should be stopped when facing a pressing epistemological urgency: in order to find something certain" (Ibid.).

Where will Descartes search for this something certain? First of all, in divine vision, "in the existence of God, as the author of the thoughts of those who mediate" (Ibid., p. 34). He had already found certainty in the expression *je pense, donc je suis*, and this affirmation represents a certainty because "he who mediates will test with doubt the consistency of his preconceived opinions, to find something 'sure and firm'" (Ibid., p. 37). Therefore, "the only certainty so far available, expressed with various formulas with equivalent meaning: I am something, I know I exist, I am a thinking thing, I am a real thing, I am, I exist" (Ibid., p. 38).

But is the certainty that Descartes is looking for and knows he with not to find the same, in different shapes and contexts, as the modern national borders, which are paradoxically uncertain due to their excessive immanent, pragmatic certainty?

Borders liquid as wax

At this point, as far as the possibility of including sensitive data, subject to the senses, as certain, is concerned, Descartes offers a paradigmatic example, a useful bridge to understand the already-cited paradox of Modernity on national borders. The question is decisive, once again: modern boundaries are more or less certain than the boundaries in any systemic imperial context? And, is it possible to connect that "liquidity" (Aubert, 2008) to the current one (Colombo, 2014, p. 47; Kaplan, 1994)? If Modernity opened and started a process of gradual advancement of uncertainty in the geographic field, this also – and necessarily – concerned the boundaries of the emerging Nation States, or are they substantially exempt? It is clear that it is not possible to answer this question without leaving any doubts, but the question must inevitably be asked, especially in long, or very long-term analysis under the common concept of *crisis*, as already outlined.

If the dynamics of the globalization, of the imposition of economic and financial principles on a global scale, of the permeability of borders due to do the "primacy" of the economic and commercial factor on the inter-State subdivision (Arrighi & Silver, 2001) are considered, is it possible, following other authors on the first Modern Age, to define the current and modern boundaries as fundamentally liquid and therefore uncertain (Kaplan, 1994)?

Apart from distinctions regarding individual cases, it is appropriate and necessary, at least for this work, to try to give a clear answer, on the basis of the available data on the conflicts between States or on the dynamics of international relations. This has been a central topic in border studies, primarily in the past 30 years. In Yeung's opinion (Yeung, 1998), the idea of a borderless world (Ohmae, 1990, 1995a, 1995b; Ó Tuathail, 1999) is just fiction, a strong concept which hides the real nature of the actual configuration of the world economic system. If some other thought that "borders are no longer of fundamental importance; territorial, ideological and issue boundaries are attenuated, unclear, and confusing" (Jowitt, 1992, p. 307, in Thrift, 2005, p. 25), Yeung states that economy is still strictly connected to the territories, and it is impossible to affirm the "end-State" in a globalized world, because they continue regulating economic activities.[19]

In his opinion, in substance,

> [T]he "borderless world" discourse must be contested, because it has caricatured the intricate and multiple relationships between capital, the State and space. First, the capitalist State continues to perform its functions in capital accumulation and to exert influence in the global political economy. Second, capital is more *territorially embedded* in places rather than having become "placeless".
>
> (Yeung, 1998, p. 303)

Now, the really fundamental question that arises is: this indefiniteness, the permeability of boundaries, which dominated Modernity, as many authors have pointed out, does it really have its roots in the appearance of the geography of uncertainty that here is meant to start with the great geographical discoveries of the Early Modern period?

But most importantly: if "the advent of the sovereign national state gave a clearer significance to the boundaries that marked off the limits of the territory" (Gottmann, 1973, p. 135), why connect the definition of uncertain with a moment of clear definition of the boundaries? Does this not seem to be an apparently irresolvable paradox?

Actually, it is worthwhile reporting here some basic observations in order to clear up any possible doubts in this matter, starting precisely from the historical definitions of border in the most important historical passages, in order to understand why many scholars called the borders in the Modern Age *fluid* and *uncertain*.

Descartes' reflections on wax seem to be overlapped to the ones about the modern borders. Which is more important: sensitive data or the system that produced it which is at its bases? It is, after all, the same question asked by Descartes, when he inquired if certainty was to be determined by the perception of the object of a specific matter. The possible answers would follow two main lines of reasoning, after exposing some definitions and a short evolution of boundaries: one, more conceptual, moved from Descartes' ideas on reality and the dissolving of a concrete world as wax, and aims at "philosophically" linking that concept to the modern boundaries; the other is based on the moment of *crisis* which marked the Modern Age.

The logic of the borders of the past, connected to the religious and divine dynamics as in the case of classical Greece, seemed to be much stronger compared to Modernity: Plato, in his "agrarian laws" (see Platone, *Laws*, VIII, 843cd),[20] inflicts penalties of a religious nature to those who offend the limits imposed by Zeus Horios, since the establishment of precise boundaries was a divine guarantee which conserved substantial social order, in the relationships between *poleis*.[21]

The boundary is the element that seals a political and religious idea, strongly linked to territory, although not in a static sense, since the very concept of empire does not allow restrictions of any kind, neither temporal (it is considered eternal) nor spatial (it is unlimited): "the basic idea of self-representation that the Romans have of their empire is that of a boundless empire, an empire that literally coincides with the world, whose boundaries are the boundaries of the world" (Schiavone, 2008, p. 29). The borders of the empire therefore coincide with the borders of the world, and vice versa.

If, in the example of Rome, the drawing of the boundaries delimits an identity (political and legal), established in the same concrete gesture, with Alexander the Great, the identification of Hellenism would be strengthened through the act of crossing the borders which, according to some scholars (see Chabod, 2007), coincided with the emergence of a first European identity, established in relation to an otherness placed outside the borders of their own civilization, their own way of being and communal sense.

The transition from the Roman Age to the Middle Ages occurred through a process of shifting of the European axis from Rome to the barbarians, symbolized in the shift of the geographical, political and – in a broad sense – "cultural" centre from Rome to continental Europe. The world hitherto was geographically and culturally limited to the European context, beyond which monstrous entities were thought to exist; the unknown aroused terrifying fantasies in the minds of men at the time: medieval maps, not surprisingly, marked a boundary of "death" around the known world (limited to Europe, Africa and Asia).[22]

The historic transition to the Modern Age was determined, instead, by the axial shift from Europe to the *Mundus Novus* and, through it, from the Mediterranean of the European inter-State trades to the Atlantic as a highway for the movement and the flow of people, ideas, cultures and resources which, at the same time, were tangible and intangible assets: from the *Non plus ultra* to the formulation of the *Plus ultra*, as Charles V wanted to establish in the coats of arms and heraldic emblems, to signify the overcoming of the human boundary of the Mediterranean, an achievement that brought with it conflict and further problems for Europe at the time.

The inner limits of the Old Continent were overcome and, with them, the same medieval mentality, but the European occupancy of the New World meant another race for the establishment of new boundaries, in a highly controversial and certainly contradictory dialectic within Europe, at the same time uncertain compared to the Western frontier (established by the Strait of Gibraltar) and its internal borders (Cantù, 2007).

On the one hand, the overcoming of the European medieval boundaries peculiar to the Old World beyond the Pillars of Hercules was not only physical and geographical but also cultural and philosophical; on the other hand, the race to the new lands, their conquest, passed first of all through the establishment of *other* boundaries at the expense of indigenous peoples and through the comparison of balances of European powers and the establishment of new spheres of State authority.

The measuring of the World, which occurred more or less systematically with the introduction and the birth of scientific geography, was a fundamental step in the process of taking possession of the Earth (see Sloterdijk, 2013). It was the signal and guarantee of control over it, an essential element for the placing of borders and demarcation of zones of possession. It mainly was a sign of certainty, underlined by the borders through tangible signs, clear elements which were obvious to the human eye, which would not give rise to uncertainties or misunderstandings (Zanini, 2000).[23]

In the words of Paul de Lapradelle, also cited by Gottmann in his essay, "the problem of the modern frontier, a delimitation of equal jurisdictions, is inconceivable in the imperialistic stage of a powerful and solitary State". Furthermore:

> [N]either the Roman Empire nor the Frankish Empire had any conception of the modern boundary. The *Limes imperii* is not the result of an agreement, even an imposed agreement, but a mere voluntary halting place . . . a conception of the modern boundary, presupposing equal jurisdictions, could be applied only in the interior of the imperial framework, and not at its periphery.
> (de Lapradelle, in Gottmann, 1973, p. 135)

This is however, the sign of inherent uncertainty in the modern condition and of its intrinsic contradiction, remarked by Wust with the image of a vortex of knowledge that paradoxically leads to a condition of uncertainty.

In the specific case of borders, with the need to overcome their "immanence", I will try to find their underlying logic, to go beyond the simple objective facts, transposing Descartes' wax metaphor onto the boundaries in Modernity which were also vulnerable to constant change and to their "fluidity".[24]

If it is true that the "fluidity" of modern boundaries in the geographical, political theatre of Europe after the great discoveries, the more the boundaries are drawn clearly, with visible fortifications, where the extension of "no man's land" is more remarkable, all the more, then, will be the political crisis between the two, be acute, reflecting a state of substantial mutual fear (Zanini, 2000). This was the case – absolutely emblematic – of the borders in some areas of Europe.

When dealing with the perception of objective facts, Descartes gives an example of a piece of wax. He basically says, citing the clear words used by Urbani Ulivi, that a piece of wax offers sensations of smell, of touch, of sight, but, "approached to fire and melted, it loses the previous sensory characteristics, while remaining wax" (Urbani Ulivi, 2001, p. 40). What happens to the boundaries of Nation States, in the long term, seems to be very similar to what happens to Descartes' wax.

The simple, pure sensory factor, which also, of course, must be considered, is not of interest here: what underlies it appears more interesting, what it has produced and surrounds it.

Remaining with the wax metaphor which seems valid for symbolic data of the boundaries, it can be said that "the term 'wax' does not mean the sensibility data. It implies something else" (Ibid., p. 40), exactly as borders should not be understood, at least here, as only as territorial subdivisions, but must be considered in their organic unity, as products of a culture, an era, a way of thinking. Furthermore, it is necessary to analyse them in the long term.[25]

Therefore, if the result of those borders, in the very long period, is the permeability that we experience today in the current borders, it is essential to reconsider what apparently may seem the most certain thing, at a given moment. That is, to put it philosophically as did Descartes, we must "continue in the observation of that piece of wax that still seems to have those imaginative qualities, such as flexibility, changeability and extension", which is opposed to the logic of Aristotelian man, who, "confident in the reliability of data of senses and imagination, based his knowledge on perception, according to the famous formula *omnis notitia humana a sensibus surgit*" (Ibid.). But Descartes, having "removed perception, discovers that what makes any knowledge possible, even sensory knowledge, is pure intellectual understanding", which is equivalent, "in terms of epistemological foundation, to the *cogito*, which was the irremovable remnant of the subjectivity of the knower" (Ibid., p. 41).

And finally, to conclude the parallelism between the Cartesian wax and the boundaries of States, it is possible to cite a clear expression in this regard offered again by Urbani Ulivi: "with the example of the wax, Descartes has not shown that the sensory qualities in the liquid wax are gone, but that they are unstable" (Ibid.), in exactly the same way as, considering the long period from Modernity to "post-Modernity", as the boundaries placed between the Nation States seemed to be: in permanent transformation, in an oscillatory state of continuous liquefaction and reinforcement, configuring a condition of political collective uncertainty.[26]

This uncertainty, deriving from the modern context of *Jus Publicum Europaeum*, of European internal conflict and deep crisis, with the disarrangement of its political structures and of the categorical structures used to understand them, was found precisely in the conceptual and factual elaboration of the boundaries themselves. We are living today – from this point of view – in the long wave of borders in the Modern Era, that, at the beginning of the 16th-century, were defined as liquid and uncertain: this would be the answer to the question of uncertainty of the modern boundaries and their progressive liquefaction, as we have defined it.[27]

It would seem that the wax metaphor can be well-applied to the idea of modern boundaries: they were signed properly, but the long-time process of capitalism and globalization produced a continuous liquefaction which began during the Modern Age (see Agnew, 2009). It is not a case that the wax metaphor was proposed by Descartes, one the most important philosophers of the Modern Age who understood and, in some way, explained the uncertain nature of the modern condition.

Considering not only the real, tangible facts but also – more than that – the symbolic and metaphorical facts which are under political logics or relations between States, the boundaries in an imperial context would make up a state of "geography of certainty" in an imperial context, as the same underlying imperial logics seem to refer to a mental, dynamic framework containing certain, fixed, transcendental presuppositions that tend to provide certainty.

The imperial borders do not, at all, consider the possibility of their limitations, and in this, they find their intrinsic certainty, while in the 16th-century, the State limitations were facing, and had to cope with, a situation of endemic and ongoing geographic and political instability, up to the current situation, in which "the boundaries have seen their functions considerably reduced, although all the same functions continue to operate in principle" (Gottmann, 1973, pp. 135–136).

This paradox occurred precisely during the European crisis of the 16th-century (Parker & Smith, 1978; Trevor-Roper, 1967), although in different forms and for very different reasons from the current ones, even if a single common thread would seem to connect them. And this is our second answer – the more concrete one – to the questions posed earlier, strictly connected to the idea of the wax.

The boundaries are, indeed, the factual consequences of the condition of mankind in the political divisions of the world: if that condition of *crisis*, marked by fluidity – as Marx, the authors of the School of Chicago in the early 2000s, Berman (1988), Beck (1992) and Bauman (2000) have very correctly pointed out – the same fluidity marks the boundaries between States, basically because "people may act under conditions that they might not otherwise choose, but they do act. They are agents who develop ideas, test them out, and sometimes, through them, give sharp impetus to transformation over both time and space" (Cox, 2002, p. 366).

The comparison suggested here seems to be on common ground with the *general crises*[28] which marked the two periods of history, as we will better analyse in the last chapter. Indeed, in a period of *crisis*, the most common problem is the confusion between national and international scales, being a "state of emergency" (Schmitt, 2005; Colombo, 2022) or a "state of exception" (Agamben, 2005).

Indeed, "the crisis can confuse or totally crush the distinction between international order and interior order, . . . because some things always slip away from this distinction" (Colombo, 2014, p. 57). The crisis contributes to confuse, or mix up the two different scales: the national and the international one. The international crisis then becomes an interior crisis, and this is much more evident, clearer, in a situation such as the modern European one has been and is at moment.

The two periods of crisis combine together, as we will see, and are comparable due to the same critical situation, which contributes in confusing both the international crisis and the interior one. In such a situation, the boundaries seem to act as Descartes suggests. As the qualities of wax are "unstable", it is possible to state the same for the qualities of modern boundaries, because they were unstable in a *crisis* (Parker & Smith, 1978; Trevor-Roper, 1967), revolution and conflict that dominated Europe during the Modern Age (Tilly, 1993). In a situation such as this, in which different authorities were emerging, in which there was no longer a fixed centre of the world, as there was in the Middle Ages (Gurevich, 1985), the boundaries seemed to act like Descartes' wax: they *were* fluid and uncertain (Aubert, 2008) like wax, and they *are* now fluid and uncertain (Colombo, 2014) like wax.

Notes

1 As Maria Antonietta Falchi Pellegrini pointed out: "the Empire, which has lost his political weight, even conserving moral authority, . . . is composed by kingdoms, towns, corporations with own structures, laws, habits, which no more recognize its supremacy" (Falchi Pellegrini, 1981, p. 17).
2 The same Gabriele Pedullà clarifies the terms even more explicitly when he outlines that "Machiavelli aims to destabilize his reader, increasing his insecurities" (Pedullà, 2013, p. XIII).
3 See Ricci, 2015a; 2016.
4 Federico Chabod emphasizes this decline of medieval certainties and principles, stating that: "nobles and plebeians impoverished in the narrow, calculating deceitfulness, in the fragmentary dispute, where the gravity of a resolute purpose, and even the formal greatness of the personal heroism are gone" (Chabod, 1964, p. 48).
5 In order to express the universal tendencies of medieval times, Gurevich outlined that in the society, during the Middle Ages: "the graphic arts were almost entirely anonymous in the Middle Ages. But even if the artist was known, it is still true that he saw as his main task the reproduction of the traditional and unchanging modes, and the expression of commonplace notions and ideas" (Gurevich, 1985, p. 37).
6 Machiavelli essentially traces, still using the words of Chabod, "the rapid but formidable map of Italian history, in its latest results, contemplated from the Renaissance" (Chabod, 1964, p. 20).
7 Indeed, about the society aspect of the medieval way of life, even the Church organizing structures was deeply inserted in the worldly social fabric, and their function is understandable only in connection with this (Brunner, 1980, p. 42).
8 While considering to a lesser extent the strictly geographic factor, the Secretary started precisely looking at the reality of his time, then arriving to a vision that Gramsci has defined of "utopian nature" (Gramsci, 1949, p. 3).
9 We can say that the territory returns to be the fulcrum around which all must revolve, because, like Turco states, "it is the condition of human action, a configuration of the world, and more particularly of the Earth surface, which allows the full development of human work" (Turco, 2002, p. 9). We understand even more clearly what has been said

Early Modern European political geography and uncertainty 67

so far if we consider one of the more explanatory definitions of political geography, provided by John Agnew: "Political geography is about the geographical distribution of power, how it concentrates and how it shifts between places over time" (Agnew, 2002, p. IX). This definition closes the circle on our reflection on the Prince. A Prince who will act on a political reality radically changed and who will – inevitably – collide and come to terms with the territorial reality, with geography, to pursue those (collective) aims always cited when dealing with this crucial work of Machiavelli. Geography is unseen, in *The Prince*, but it is always present, and it is the essential basis for an organic study of the political dynamics, in the 16th-century of Machiavelli as in the present day.

10 This is what happened – in an audacious historical parallelism proposed by Giacon in 1950 – in the aftermath of World War II, when the authors of the second Scholasticism were resumed: "an intense and insistent effort is indeed ongoing to lay the foundations of a national and international associated life, in which allow the conditions for the fulfilment of the unstoppable desires for justice and peace pursued by everyone, the need is strongly felt for a return to doctrines that, after having passed the exam of centuries, might offer safer and stronger principles than the ones given by more recent, but deceptive and ruinous doctrines" (Giacon, 1950, p. 8).

11 An indeed "if the ambitions of princes and rulers were mainly directed to enlargement and success, there must have been inevitably not only those who would be irritated in the practical order, but also those wondering if the concern of princes and rulers was theoretically legitimate" (Giacon, 1950, pp. 22–23).

12 The author observes that the difference between the empires contemporary to Bodin and the imperial power of the time of Charlemagne can be summarized, in essence, as a "lack of power".

13 Indeed, using the words of Isnardi Parente, "after the decline of the Empire from true kingdom to principality, for the arrogance of the German princes, which have reduced its power to a shadow, only a claim of universal dominion remains of the memory of this royal power, sovereign of a vast territory, not an ideal value, but a vain shadow of power" (Isnardi Parente, 1964, p. 59).

14 And yet, at the same time, "Bodin feels that he can read, through the historical events, the story of the great natural, celestial events that determine it to a large extent, as in a cosmic sequence" (Isnardi Parente, 1964, p. 83).

15 With Botero "the 16th-century closes, the Middle Ages and the autumn of the French Middle Ages dies, the French humanism dies" (Isnardi Parente, 1964, pp. 99–100).

16 At the same time, it is also true that we are in front of what may seem like another huge, at least philosophical, dilemma, because the work of Descartes has among its targets the "certain and evident cognitions".

17 "The clear and distinct evidence will always establish for Descartes the ultimate (or, rather, the first) criterion of the truth" (Urbani Ulivi, 2001, p. 27).

18 Such doubt on the certainties of the world also involves those who believe in God, for whom "remains open the task of the epistemological foundation, because the total devaluation of human knowledge it is compatible with the belief in God", such that "the sceptical challenge is not only valuable in an atheistic perspective, but also within a religious viewpoint", and therefore "the uncertainty of knowledge would remain . . . even from the point of view of an atheist" (Urbani Ulivi, 2001, p. 29).

19 "Territorial differences and geographical unevenness remain integral to globalizing processes. The world is made up of a dynamic mosaic of uneven geographical formations that are shifting over time, subject to the interplay of power relations among the state, capital and other forms of social institutions. The analytic logic of a 'borderless' world becomes absurd in this. Contesting the borderless world interpretation of ongoing changes in the global economy, because by portraying an end-state, it fails to recognize its dynamic transformations over time and space" (Yeung, 1998, pp. 303–304).

20 Quoted in Zanini (2000, p. 41).

21 "When the linear border has formed, also and at the same time, the limits of the territory of another State, it has ensured the stability of the current system and safeguarded the need to implement the freedom and autonomy of the polis" (Daverio Rocchi, 1987, p. 27).
22 The word *M-O-R-S* is distinctly written at the four points around the Hereford *mappamundi*, encircling the globe (see Fig. 5.2). In this cultural context, the Pillars of Hercules constituted the natural and symbolical passage over which the deity does not "guarantee" survival, the point that the European man must not pass: hence the famous formula, *Non plus ultra* (see Gregory, 2008).
23 It is clear that, with the rise of Nation States and of the consequent newly born European system, the border of subdivision of powers and internal equilibrium imposed by the multiplicity of legitimate competitor authorities might firstly seem much more "certain" than the imperial *limes*, than the demarcation – that could be called as indefinite – of imperial logic that, as such, does not admit borders at all.
24 If we continue to give value to the interpretation that assumes as unquestionable the fact of a strong and effective economic globalization, but not corresponding – in terms of conceptual extension – to a political globalization (therefore without "coextension" between these two areas, see Colombo, 2010), surely the idea of a trespassing, of a continuous overcoming of State borders by economic agents and objects of the financial economy seems to be entirely consistent with the current times.
25 It is true that a certain vagueness characterized the configuration and the spatial structure of the Holy Roman Empire – as well as any actual *imperial* organization – but at the same time, it is necessary to understand this "mobile" structure not in its given – or not given – form, but in its conception. Meaning with this that the modern uncertainty of the European internal structures of State demarcation is not so much due to their "immanent", so to speak, presence, but much more to their symbolic and conceptual value.
26 In theory, in a capitalist economic system, within the Western world system – with the exception, therefore, of the special cases of territorial contexts of conflict or tension between neighbouring countries – such as the two Koreas, paradigmatically configured, it is possible to answer that we surely live in a generalized context of uncertainty about borders and their substantial permeability, in a *fil rouge* with the modern situation.
27 Perhaps because the philosophy underlying the boundaries, it is stronger than the tangible factor in itself – the definite border placed between a State and another – even considering that "a border, however, does not guarantee to be completely impenetrable" (Zanini, 2000, p. 50).
28 I will better define this concept in the conclusive chapter.

4 The "mad flight" and the geography of uncertainty

The accomplishment of the "mad flight"

We have so far seen which Modern Era theoretical and philosophical presuppositions lead back to uncertainty. It is the aim of this chapter to travel from a fascinating literary expedient, that of the "mad flight" (in Italian "*Il folle volo*") from Dante Alighieri's *Divina Commedia*, XXVI *Chant* of the *Inferno*, to reach and whence apply the same idea of uncertainty to the "mad flights" of the explorers and travellers who gave rise to the Modern Age. Thus, showing how geographical uncertainty coincides with the great discoveries of the beginning of the 16th-century.

This same concept of uncertainty will be linked to Europe opening up to "new worlds". We will attempt to unearth similar cultural roots and concepts in Dante's mad flight and link them to those which can already be seen at the start of the Modern Age as suggested by Piero Boitani (1992). By going beyond simply quoting Dante, we can clearly see how the end of Ulysses' journey towards the unknown, the impossible, is comparable to European exploration travels in the 15th and 16th centuries which led to knowledge – and later to the awareness of this knowledge – of the "new" continent, the land that Amerigo Vespucci defined *Mundus Novus*.

The exploration travels of the Modern Age contributed to radically change the European way of interpreting, understanding, experiencing and representing the world. They produced an intellectual shock, a global revolution in many ways, above all politically and geographically. I will deal with these aspects later.

This Modern Age crisis mentioned and visible in 15th- and 16th-century Europe gave way to the opening up of the Old Continent to new spaces, to a "New World". We must start from here, from these geographical coordinates, to fully understand this uncertainty and its tie to the Modern Age and to geography. It cannot be separated from the dynamics of travel, from the mad flights of exploration which are fundamental: they are the bearing axis founded on uncertainty.

The "mad flight" is *geography* and *uncertainty*

In "madness" and "travel", we can, clearly and congruously, trace some of the key concepts that we want to address here in fitting, poetic form. The idea of a "mad flight" certainly includes, on one hand, the "madness" animating the spirits of the

DOI: 10.4324/9781003394204-4

explorers of the early Modern Age and, on the other hand, the "flight", the journey itself, the adventure and the exploration which distinguished the Modern Age revolution. It was, not by chance, above all, a geographical revolution (Schmitt, 2015).

From a more theoretical perspective, it would seem possible to compare our statements to some reasonings expressed by Emanuela Casti (1998), when she deals with cartography in the Modern Age: according to Casti, "geography in full sail" began to take shape thanks to the renewed force of knowledge in the Modern Age and it represented the way the great explorers learnt about the world, the explorers who abandoned all material certainty of family roots and affections because of a yearning for knowledge, a thirst for knowledge and craved discovering of the world. An aim which, of course, in many cases, also had utilitarian premises, as in the case of Columbus, for example.

The historian Felipe Fernández-Armesto reasons about it in rather harsh tones: "what mattered to Columbus was not so much where he was going as whether, in a social sense, he would arrive" (Fernández-Armesto, 2009, p. 99) because "'to be a great lord' was Columbus's driving force" (Ibid., p. 102). Certainly, this point of view is a partial one, since personal and social motivations cannot be univocally analysed. Furthermore, in the most important cases such as Columbus', which can be considered one of the most striking, an amount of luck, randomness and ruthlessness, as the scholar states, perhaps exaggerating, must also be added together with a certain vision of the world, corroborated by studies or by deep personal convictions so as to animate the choices of the traveller. The more theoretically revolutionary component played a vital role. It had rooted beliefs of how the world should be, how big it should be, what the laws that governed its life were and so on. Interpreting the concept used by Casti, this was essentially "geography in full sail".

In a broader discourse concerning the function of the map in Modern Times, this way of doing geography is accompanied by another very suggestive concept: "laboratory geography" (Casti, 1998, pp. 22–23). According to Casti, these two approaches to the way the world is examined are complementary, and acquire prestige in the Modern Age in spite of being considered different and "separate moments" of a general understanding of the world's features and borders. Although sometimes, necessarily, "geography in full sail" coincided with the voyages of discovery and the factual knowledge of the world, the knowledge acquired by the explorers was essential to "laboratory geography", to the mapmakers, who transposed on a plane the knowledge acquired by travellers and vice versa the maps were used by those travellers to face the open seas, in a game of continuous references, revisions, delineation of continental boundaries and adjustments: "the map becomes the *trait d'union* between these two ways of investigating the world and conceiving space. Thanks to cartography the *globus mundi* becomes a *globus intellectualis*" (Ibid., p. 23). Furthermore, Casti quotes Éric Dardel who adds,

> [W]e can talk about the poetics of geographic discovery, in as much as the discovery was the fulfillment of a sight which included the whole world. It was creation, the creation of space, opening up to the world, to an extension

of the human being, an impulse toward the future and it defined a new way of establishing the relationship between man and Earth.

(Dardel, in Casti, 1998, p. 22)

Understanding what is meant by "mad flight" in geography is fundamental in such a complex context which extends from geography and knowledge to other fields.

The journey of Ulysses is considered by Dante "mad" because it is directed towards the unknown, towards that which cannot be included within the perception of *certain* and *sure*, and moreover, because he is not led by God's hand, but rather, is driven by a personal need for knowledge: it is not a journey undertaken in the light of divine will, such as Dante's in the *Divine Comedy*. The two journeys are in contrast: "Dante's journey differs from Ulysses' travel precisely because it does not embody the madness, it is wise and sublime in its Christian concept. It is orthodox and integral" (Boitani, 1992, p. 63).

The "madness" that leads Ulysses is therefore pagan, far from the late Middle Age catholic ideal that Dante interprets, and that drives him, like other heroes, to go beyond a boundary that leads him into a world unknown and full of dangers, because it is a "world that hath no people" (Alighieri, *Inf.* XXVI, 117). In effect:

[T]he adventure will end in death but only when the hero reaches the point where he can see what is not-visible: The mountain in Purgatory, more specifically when the itinerary leads the hero out of the knowledge area of Nature.

(Azzari, 2012, pp. 32–33)

The point that Margherita Azzari highlights in this short paragraph is essential when considering the madness of the flight of Ulysses, because it is a journey towards *uncertainty*, a direct flight beyond what delimited the world of medieval certainties. That's why – in a deeply symbolic way – the fulfilment of Columbus' undertaking leads to the assimilation of a first geography of uncertainty which coincides with the Modern Age. Geographic uncertainty is the metaphorical and factual uncertainty of the madness of the flight of Ulysses, which is dictated primarily by the unknown destination and then – but not as a secondary element – by the typically pagan heroism of Odysseus.

In other words, Ulysses represents the anticipation and the incarnation of an ideal of the modern tragic hero. Precisely, the one who detaches himself from the proposed, experienced certainties of the Middle Ages and heads towards unknown regions, passing over all known conventions. He relies on his own experience and his intellect.

Boitani underlines this fundamental passage, together with other aspects. The Ulysses described by Dante is for him the

legendary hero searching for a target beyond that of the ancient classic or pagan times, beyond the ontological limits of his own culture; the late Middle Age philosopher who takes over Christian wisdom; the new Adam; the

Ulysses of the 14th century who personifies in tragedy the birth of the modern world.

(Boitani, 1992, p. 61)

Therefore, according to Boitani, Dante's Ulysses is first of all a mythical hero whose tragic story started as a search for a destination outside "Christian borders", and incarnates the tragedy of the Modern Age condition, where every certainty is lost and where the extension of European space had in itself "disoriented" man.

Tragedy, the birth of Modernity and, we may add, *uncertainty* are embodied in Dante's Ulysses. This opens up a fascinating prospective, which I will deal with more extensively later, on how the element of tragic heroism coincides with Modernity when seen from a geographical point of view at the beginning of the Modern Age, seen as a new event or a reproposal of the idea of tragedy. This thesis has been put forward by several studies in various areas (see Lombardo, 2005; Schmitt, 2006; Watt, 1996).[1]

Forerunner and pioneer of this idea of tragedy, which represents a deviation from the medieval – universal and traditional – framework is the very Ulysses proposed by Dante. The hero relies solely on his own human abilities (following the formula *virtute et canoscenza* [virtue and knowledge], a clear expression of the Aristotelian *megalopsykhia* [$\mu\varepsilon\gamma\alpha\lambda o\psi v\chi i\alpha$]), and not on metaphysical or divine elements. If Dante's Ulysses, pioneer in travelling the modern world, is the one who tries, in his "mad" flight, to go beyond the limits of the Pillars of Hercules in an authentic, ideal crossing of medieval boundaries, Christopher Columbus represents, in a certain way, the living incarnation of the one who completes the journey undertaken by Ulysses, bypassing every metaphysical and geographical certainty submitted and dictated by the Middle Ages.

Surely, in the late 16th-century, the myth of the Pillars of Hercules for the Italian poet Torquato Tasso has already been completely surpassed, thanks to, amongst others, Columbus' ventures. Indeed, he wrote in the *Jerusalem Delivered*:

> A knight of Genes shall have the hardiment/Upon this wondrous voyage first to wend,/Nor winds nor waves, that ships in sunder rent,/Nor seas unused, strange clime, or pool unkenned,/Nor other peril nor astonishment/That makes frail hearts of men to bow and bend,/Within Abilas' strait shall keep and hold/The noble spirit of this sailor bold.
>
> (XXXI)

In other words, we find in Torquato Tasso the exact idea of the fulfilment of the mad flight of Ulysses and, with the voyage of Columbus, the overcoming of the medieval belief in the impossibility of going beyond the Pillars of Hercules, "when Columbus was not only returned several times to overcome them without dying in the attempt, but he gave news of a new continent (although discovered by mistake)" (Longo, 2013, p. 66).

It is true that the Pillars of Hercules, as reported by Nicola Longo, "have been known from antiquity as free from prohibition", even if Pindar had already suggested the idea of the impossibility of crossing them (see Ibid., pp. 71–72).

These same links between Dante's Ulysses and Columbus are to be found in Boitani. The encounter between interpretation, poetry and history generates, as we shall see, a typological sequence in which Dante's Ulysses represents the "body" and Christopher Columbus the "fulfilment", in which reality "fulfils" the "scriptures" prophesying them, building a rhetoric and a myth that in one form or another the modern world sees as its own almost until the present day (Boitani, 1992, pp. 61–62). On the one hand, Ulysses, as "body", on the other hand, Columbus, as "fulfilment" of the same figure, implying direct correspondence between the two characters, which have both elements of commonality and difference. They represent two different symbols in overcoming defined limits set down by man. There is an adjective that separates these two worlds: "mad". Ulysses may be the character that anticipates Modernity, but only when it is not yet complete; Columbus, instead, represents its first icon, almost the "initiator". Both, however, are also linked by that same adjective, mad, which is so paradigmatic of the deliverance from the old medieval canons. Their journey is mad because it is uncertain in terms of destination and prospect of travel, and also because – although caused by personal beliefs – it generates in turn several aspects of "uncertainty".

This vision of uncertainty which dominates human life in its "divine yearning" is intrinsic to its own condition, and it is so, even more metaphorically, for he who embarks on a journey into the unknown, if we reconsider Wust's thesis and adapt it to the metaphor proposed here: according to Wust,

> [T]he fundamental condition of man . . . is constituted by a structural lack of security. However, the same dynamics of life, which continually require decisions to be made and positions to be taken, force the individual to project himself into the unknown, to take up and travel along paths without knowing their final destination.
>
> (Wust, 1985, pp. 18–19)

This then seems to be the perfect tally of the journey of Dante's Ulysses into the unknown, a journey that is the very symbol of human life but even more is the sign of modern uncertainty which originated precisely in the geographical field.

It is essential, in this vision provided by Wust, to bear in mind that "pure intellectual and philosophical knowledge is therefore inadequate to be able to penetrate the multiple aspects of existence" (Ibid., p. 19).

It is no coincidence that the German philosopher chooses one of the best-known parables of the Bible to express the necessity of uncertainty for some sort of ascension towards the divine, that day-to-day certainties could not contemplate. His work on uncertainty opens with the parable of the "Prodigal Son", which is also, in some ways, comparable to the story of Ulysses, at least in the uncertainty of the destination and of the journey itself, which is unknown, but aspires to the highest peaks offered by the world of certainties.

Why the "geography of uncertainty"?

That significant conception of the globality of the world, even if it was initially a conviction which fully emerged only at the beginning of the 16th-century, opened up the Modern Age (Marramao, 2012). To put it in literary terms, it constituted the accomplishment of the "mad flight", briefly mentioned earlier, came about thanks to the "geography in full sail" mentioned by Casti.

To get to the heart of the problem and address the point in different terms: how is it possible to connect the onset of European man to oceanic, global spaces and the affirmation of "global linear thinking" (Schmitt, 2003, p. 87) to the idea of uncertainty applied to the geographical field? Is it possible, in this context, to speak of the beginning of a "geography of uncertainty" as a loss of the certainties that had characterized the medieval European world?

The beginning of the Modern Age coincided, above all, with a revolution of the geographical perspective. This revolution was based on territorial factors tied to the travels, in particular to those travels defined more broadly as mad flights of the explorers when compared to the ventures of Ulysses. The explorers, starting from Columbus, conveyed the mentality of European man at that historic moment in a paradigmatic and practical way. It was a profound transformation in his mindset of knowledge, because of the "global" discovery of the world. A world that, until then, had coincided with Europe itself, so that the concept of civilization directly collimated with the European one, without any possible exception in the medieval mindset. The discovery of a New World, which sprang from an uncertainty of background, led to a radical reinterpretation of the cultural, philosophical, religious and economic overview and of political authority. Due to the extent of these modifications, it is in fact possible to speak of a revolution of ideas, a new era, and to start talking about a geography of uncertainty.

The words that Amerigo Vespucci wrote to Lorenzo di Pier Francesco de' Medici perfectly befits this concept of uncertainty. It dominates the travels of exploration and has an impact on other areas. This can be, therefore, considered fundamental for the same perception of the New World, a sort of mark given to the Early Modern period and to what might be called a first, real form of globalization. Vespucci, after stopping at the "Fortunate Isles", tells us that "having there recovered our strength and taken on all that our voyage required, we weighed anchor and made sail. And directing our course over the vast ocean toward the Antarctic we for a time bent westward". From there, he sailed with his fleet for two months and three days, and "any land appeared to us". And here is the part that most concerns us in this case:

> I leave to the judgment of those who out of rich experience have well learned what it is to seek the uncertain and to attempt discoveries even though ignorant. And that in a word I may briefly narrate all, you must know that of the sixty-seven days of our sailing we had forty-four of constant rain, thunder and lightning – so dark that never did we see sun by day or fair sky by night.
> (Vespucci, 1916, p. 2)

"To seek the uncertain" is the essence of the travels of Vespucci who clarifies what we have addressed earlier. A geographic uncertainty, concerning both the trip and the state of mind – anxiety, discomfort and danger – dominated those who unfurled the sails to the winds on unfamiliar routes, the interpreters of a geography in full sail.

They departed in almost total uncertainty, although aware of their destination, the itinerary and the routes. Uncertainty, shaping their very souls, was the underlying reason for their travels. An uncertainty, however, which still tended to revolve around religious ideals and sometimes with missionary ideals of that same journey, so that Vespucci wrote: "God's mercy shone upon us much when we landed at that spot, for there had come a shortage of fire-wood and water, and in a few days we might have ended our lives at sea. To Him be honor, glory, and thanksgiving" (Ibid., p. 3) and "we anchored off the shores of those parts, thanking our God with formal ceremonial and with the celebration of a choral mass" (Ibid.). So the journey, the enterprise, even if dominated by uncertainty and by the confidence in their own personal abilities ("because I have ever delighted in virtuous things" (Ibid., p. 58), was still strongly imbued with religiousness, with a mystical and transcendental vision observable first and foremost in Columbus.

It is interesting to note the reference made by the same Vespucci to the XXVI canto of the *Inferno*, going beyond the vision of Dante and contradicting the great poet, who said that the world outside the Pillars of Hercules was a "world unpeopled". Vespucci in the *Letter to Soderini* (see Salvatori, 2013) writes that

> I read it was held that this Ocean Sea was unpeopled, and our own poet Dante was of this opinvion, when in the twenty-sixth chapter of the *Inferno*, he depicts the death of Ulysses. On this voyage I saw things of great wonder, as Your Magnificence shall hear.
>
> (Vespucci, 1992, p. 59)

The reference to Dante's Ulysses by Vespucci again confirms the connection between the two areas and the accomplishment of his mad flight by him, Columbus and other contemporary great explorers. According to Boitani, "Vespucci considered geography 'real', but Dante considered it fictitious, mythical and existential. He corrects it in the light of its own 'discoveries'". But above all, "the route of Vespucci is the same as Dante's Ulysses. The Florentine navigator knows it and wants it", going so far as to "interpret his own travels in the light of those of Dante's Divine Comedy" because he "loves to see himself both as a Ulysses and a Dante" (Boitani, 1992, p. 72).

Essentially, we can find the myths of the past in the great deeds carried out by Columbus and then by Vespucci who updated and embodied them in a bond that finds its meeting point in what has been called "the mad flight". In other words, "the poetry of the past already contains the modern world" (Ibid., p. 73), so that in 1892, Gaspare Finali – quoted by Boitani – "conceived the idea that the great Christopher was in fact inspired by the Dante's story" (Ibid., p. 74), and

> the comparison with the final journey of Dante's Ulysses is almost automatic, so naturally appropriate, so perfectly emblematic of the encounter between poetry and history over time, to appear obvious even to us, sharp

post-figurative and post-modern supporters of the division between reality and fiction.

(Ibid., p. 73)

The European perception of a "full" globality of the world and the awareness of the existence of a fourth continent coincided at the same time with the possibility of extending the reach of European action in politics, religion, commerce and other areas. It is the European space itself which is able to find new territories and "spaces of action". These extended not only beyond the simple ideal perspective but also in pragmatic and absolutely concrete ways thus contributing to substantially modifying the point of view (and the certainties) of Europeans. In the 16th-century, full awareness was acquired of being able to transpose the logics underlying the conflicts and attempts for supremacy onto American soil, as European internal conflicts for power, for religious and confessional reasons, characterized Europe at that time.

The uncertainty of new world geography

This radical, revolutionary and profound change in perspective, which had finally become global, is evident and is clearly visible especially in cartography, in the ideas, concept and cartographic signs and visions associated with the discovery of the New World thanks to new interpretations of a worldview arising in the early 16th-century, which constitutes a further field of investigation of the geography of uncertainty.

With the enterprise of Columbus, the whole world becomes "new": subject to the geographical image, it gains in its entirety a completely different meaning, an unprecedented existence and consistency, a fresh nature and a new reason. Since Columbus' travels, a reversal of the patterns of the past had already occurred, which would acquire full awareness with subsequent expeditions. Cartography acquired an increasingly scientific *status* starting directly from the Modern Age, as representations were no longer merely tied to a metaphysical vision that led the reader to decipher the map according to a religious and "sure" interpretation, but they were now based "only" on secular elements. This occurred because Columbus did the opposite of his rivals: he showed that is the Earth to work on the model and not the model on the Earth (Farinelli, 2009, p. 140).

The discovery of a New World suddenly subverted the beliefs of the past, also leading European man to formulate several hypotheses on the new configuration of the world that the great voyages were contributing to delineate, unable to complete the new maps in the affirmation of the "*terrae incognitae*" formula for the unknown lands (see Lois, 2018). Some scholars were still dominated by "a personal feeling of uncertainty about the actual meaning attributed by the authors of the time to terms such as 'terra firma', 'discovery', 'New World', 'Asia', 'continent', and all the other expressions which flourished in the age of discoveries" (Washburn, 2007, p. 26). For centuries, including the last one, no conceptual agreements had been found on the terms quoted earlier, having man still to define the globe in a systematic and comprehensive way.

According to John H. Elliott, it was "reasonable enough that there should have been continuing uncertainty throughout the sixteenth-century as to whether or not America formed part of Asia" (Elliott, 1970, p. 40).

The turmoil led by the great discoveries of the 15th and 16th centuries was also reproduced in the terminology to be used in the geographic field:

> [W]hile the maps accompanying the text [of Waldseemüller, *Cosmographiae introductio*] seem to suggest a new hemisphere and a new continent (although in one case separating North and South America and not in the other), the westward extension of the "New World" is problematic and indefinite.
>
> (Washburn, 2007, p. 27)

Furthermore, as far as geographical terminology is concerned, "the translation of 'terra firma' with 'continent' in the 20th century presumes an identification with the continents as they are known today, but it is not, I think, necessarily valid for the Age of Discovery" (Ibid., p. 35).

The divulgation of the new cartographic works, vague sketches, with doubts about the outlines of the new continent, the coastlines of South America, allowed the acquisition of a far more realistic perception of the world than in the past, especially in the awareness that beyond the Pillars of Hercules an ocean opened out, with a new continent beyond it. The certainty that had prevailed for man up to that time (obviously European man, in a broad cultural and social meaning), that the world was centred on Europe, and trade markets mainly intertwined in the Mediterranean basin, was gone.

The expansion of European regional territory also coincided with a different perception on the part of the European man of his area of origin and sea of reference which had been the Mediterranean until the discovery of a New World. The pivot of medieval dynamics had become marginal. This perception was intensified by the shift of European trading northwards, with the rise of the English and Dutch powers[2]: the centre of gravity within Europe, according to Fernand Braudel,

> shifted from the south to the north, first to Antwerp and then to Amsterdam, and not – let me point out – to Seville or Lisbon, the centers of the Spanish and Portuguese empires. Thus it is possible to lay a piece of tracing paper over the historical map of the world and draw a rough outline of the world-economies to be found during any given period.
>
> (Braudel, 1979, p. 84)

The new discoveries changed the commercial European reference axis. That very same system, which had had need of a central economic and cultural pivot, changed radically and substantially by shifting towards the Atlantic.[3]

As Violet Barbour pointed out,

> geographically, early modern capitalism extended its radius, especially in the penetration of northern and eastern Europe, and in the beginning the exploitation of other continents to which the age of discovery had opened seaways to. In western Europe trade was slowly shifting from its ancient seats commanding the Mediterranean highway to northern and western ports facing the North Sea or the Atlantic Ocean.
>
> (Barbour, 1963, p. 11)

Almost right after the publication of the work of Vespucci, it was this same discovery of a New World to reduce Europe to an Old Continent. It was perceived as such. In this context, Braudel identifies the period starting from 1569 as an "economic renaissance" of the Mediterranean, although later it "became a secondary region and remained one for a long time to come" (Ibid., p. 87).[4]

As rightly pointed out by Schmitt, the discovery of the New World projected Europe towards a totally new conception of itself and its borders, compared to the medieval past. And in fact,

> with the appearance of a "New World", Europe became the Old World. The American continent was really a completely New World, because not even those scholars and cosmographers of antiquity and the Middle Ages, who knew the Earth was a ball and that India could not be reached on the way to the West, had any inkling of the great continent between Europe and East India.
>
> (Schmitt, 2003, pp. 86–87)

It was a substantial revolution of the European perspective, which was, above all, Eurocentric in the medieval context, and which saw Rome or Jerusalem as symbolic centres, but also more strictly as geographical centres of the world. The same Schmitt states: "when a 'New World' actually emerged, the structure of all traditional concepts of the *centre* and *age* of the Earth had to change. European scholars and cosmographers now saw a vast, formerly unknown, non-European space arise" (Ibid., p. 87).

Surely, Schmitt points out, a very strong contradiction of Europe originates here. It is true that the concepts of centre and age changed, but to acquire full awareness of these structural changes in the European conception it was necessary to wait.

Because when discoveries emerged in all their disruptive political, structural and conceptual, religious and utopian force for Europeans, they kept faith – certainly in a totally different way than in the past – to the idea of centrality of their continent, because they projected in the New World the logics underlying European power, tied as it was to the spaces of the old continent itself.

A New World was discovered, but this remained a sort of "continuation" of Europe, at least in the early centuries of European colonial projection on the new lands. But it is necessary to consider, at least on this occasion, the extent of the modifications produced by this discovery, which actually undermined all the principles and the philosophical and conceptual bases that had dominated the medieval past, because "the emerging new world did not appear as a new enemy, but as *free space*, as an area open to European occupation and expansion. For 300 years, this was a tremendous affirmation of Europe both as the centre of the Earth and as an old continent" (Ibid.). We have to outline this concept, because the discovery of the New World

> also destroyed previously held concepts of the centre and age of the Earth. It initiated an internal European struggle for this new world that, in turn, led to a new spatial order of the Earth with new divisions. Obviously, when an old world sees a new world arise it causes dialectical doubts and is no longer old in the same sense.
>
> (Ibid.)

The conscience of the world's globality, therefore, appears simply a geographical or cartographic matter, which of course it is not, but it has also very important political implications, because "the question was political from the start; it could not be dismissed as 'purely geographical'. As scientific, mathematical or technical disciplines, geography and cartography certainly are neutral. However, as every geographer knows, they can be exploited in ways both immediately relevant and highly political (Ibid., p. 88).

The really interesting fact that the German author reveals is that the political internal disputes which distinguished the Old Continent together with its geography, geographical discoveries and their relevant cartographic representations, despite their increasing "scientific" approach, were "distorted". They were even outclassed by their own scientific credibility.

In other words, the same geographical concepts were the victims of some sort of European speculation adapted to the political needs of the States (Branch, 2014). The political European fragmentation of the Modern Age, whose legitimacy would come with the Peace of Westphalia in 1648 and which saw the emergence of numerous different and often conflicting State entities, would also be transposed in the cartographic field. Therefore, maps and geographic charts held an important position which could make statements, emerging between allegories and old symbolisms as the true symbol of a new way to know the Earth. A greater "certainty of representation" would emerge, contrasting with the rising of modern uncertainty, since in the Modern Era, the increased circulation of maps of all kinds has greatly contributed to the development of a thought really capable of displaying world distances and diversities, gradually transforming an indefinite mass of ideas into images more and more precise (Palumbo, 2007, p. 185).

On the one side, a more "certainty" of the cartographic representation,[5] on the other, the uncertainty of the geographical dynamics, both from the point of view of the "departure" of modern travels and of the uncertainty that they have caused in other areas.

The cartographic symbolism of the Middle Ages offers several hints of considerable interest about the idea of certainty of the illustrated world and the uncertainty of what the world, as a direct expression of divine will, could not contemplate. The medieval maps, the *mappaemundi* (from the Latin *mappa*, cloth,[6] because they were drawn on cloths), offered allegorical elements to better understand what we are addressing here.

The world was perceived and therefore depicted in three sections with Jerusalem usually placed in the centre: two sections, equivalent in size, represented Europe and Africa; the third, larger and generally placed at the top, depicted the Asian continent. The East was not on the right, as we are used to conventionally representing it today, but at the top, at the coinciding points of the cosmic and solar element with the divine. The altars were placed in the direction of the sunrise, with God Himself arranging the world and human beings from the higher position. From there it was usually possible to glimpse the Garden of Eden that served as an example for men, in order to make the necessary comparisons between the mundane and the spiritual, the physical and the metaphysical, under the guidance of God and the concept of religion itself. Eden was placed at the apex of the maps as an *exemplum* for men on Earth. It represented "a map inside a map", signifying the desire to bring the world back to its original divine expression and fellowship with God.[7]

The medieval period was defined by a yearning and a desire for certainty in the cartographic field. The same representation was only an attempt to put the world on paper, or at least the vision of it at that time and directly connecting it to the symbols and dogmas of the Church. Certainty was everything, but not in the modern, concrete sense. That is the explanation of the *horror vacui*,[8] which was the expression of the impossibility of the void – and so of the uncertainty – for medieval men (Gurevich, 1985). Everything was definite, all determined in accordance with the religious logics, according to the will of God, and it could not be otherwise, as I will discuss in greater depth in the next chapter.

On the contrary, the earliest forms of "defined" scientific cartography, based on reality and on the experience of navigators and no longer on the religious and symbolic apparatus which had given form, essence and content – always religious – to medieval cartography, were established with modern cartographic representation. Here again, we find the study of geography and its representation in the Modern Era as dictated by the scientific method. In this context, a strong uncertainty shows itself impetuously, assimilated in the transition determined by the "mad flight" into the unknown. The "fulfilment" of this, embodied by Columbus and by the great explorers of the Early Modern period, also seems to represent the factual expression of the idea of uncertainty not admitted and acknowledged in medieval representation, which refused to conceive the representative void.

Some further considerations about the geography of uncertainty

After having traced the essential outlines of the "geography of uncertainty", what remains of the connection with Dante's "mad flight"? Does it really constitute an intrinsic conjunction, a strong, stable relationship? The issue offers different perspectives and possible crossroads.

What I wanted to highlight in this chapter is the relationship between the great voyages of exploration beginning in the Modern Age, linked to the concept of the completion of the mad flight, and the idea of uncertainty that seems to appear with Modernity in the decline of medieval myths, dogmas and – certain, incontrovertible – beliefs. The uncertainty is geographic because it arises from geography, from the first great voyages of discovery, with their indefinite destinations and routes, which were steeped in uncertainty and were its ambassadors in the conceptual field of geography and the knowledge of the world. Furthermore, cartographic representations tolerated uncertainty and put it down on paper without the highly symbolic medieval elements, with bold empty spaces represented – sometimes still containing religious allegories, although much more moderately than in the past – and with the gradual definition of the borders of the American coast in the early 16th-century.

The symbolic system was no longer the same, the references were lost, and with them the certainty of knowledge of the world weakened. The world entered inevitably in the domain of the uncertainty of the first globalization, which would find further constraints in the peculiar financial development and its subsequent phases, especially in the spaceless logics of the capitalist-financial economy that marks our world even in recent financial and economic crisis.

The geography of uncertainty, in other words, is nothing but a synonym, a different interpretation of the idea of globalization, which started – unequivocally – with the awareness of world globality, with the fulfilment of the mad flight of Columbus and the other great explorers of the Modern Age, who were able to overcome every dogmatic, metaphysical and allegorical definite limit which had led to conceive Europe and the Mediterranean as the centre of the world. A world which was only partially known and deprived of its entirety by the limits of the traditional scenario and which became global through the uncertain and "mad" travels.

That globality would assert itself almost entirely from the year of the turning point, 1492, when the mad oceanic flights opened up Europe – and the world – to an almost globality of the Earth and, with it, to a first globalization, a first "geography of uncertainty".

Notes

1 Agostino Lombardo, from the standpoint of his discipline, has particularly focused on the work of Shakespeare and his contemporaries, identifying in some characters, like Othello, Hamlet and Dr. Faust by Marlowe, the archetypes of the tragic heroes of Modernity. They were unable to read and interpret the changes of their time, reducing their life experience only to the intellectual factor, without finding more feedback and the possibility of redemption – or comprehensive answers – in the divine and spiritual bond, to which the medieval man inevitably made constant reference, existentially and culturally, finding the certainties of his life.
2 At this purpose, see Wallerstein, 2011, Vol. I; Prak, 2005.
3 "The shift in the locus of European trade from the markets of the Mediterranean to the North Atlantic overthrew a centuries-old pattern of commerce and established the basis for the predominant role of North Atlantic Europe in the era of industrialization" (Rapp, 1975, p. 499).

4 It is also true that, focusing particularly on the economic and trade aspects of a world-economy, Braudel will also say that "the Mediterranean of the sixteenth century was in itself a *Weltwirtschaft*, a world-economy, or, to use another German expression, '*eine Welt für sich*', a world unto itself" (Braudel, 1979, p. 81).
5 See the two volumes edited by the "Comitato Nazionale per le Celebrazioni del V Centenario della Scoperta dell'America" (AA.VV., 1992).
6 See Brotton (2012).
7 In this regard, see the fundamental work edited by Harley and Woodward (1987).
8 Quoting Naomi Reed Kline, "these mappae mundi [the T-O one], shared with other newly flourishing forms of medieval art a *horror vacui* wherein a seemingly endless array of images were packed into spaces that could hardly contain them" (Reed Kline, 2001, p. 220).

5 Cartographic secularization

Representation and modern tragedy

According to many authors (Schmitt, 2006; Lombardo, 2005; Watt, 1996), Modernity has been a sort of immersion in tragedy. The atmosphere of "irreversible reality" (Schmitt, 2006, p. 39) characterizing modern reality, which was reduced to a level of immanence and brought religion back to experience and subjective interpretations, is described – cloaked in a certain tragic dimension – in many works of the 16th and 17th centuries.

For Luigi Pirandello, the substantial difference between the tragic dimension of ancient Greece and the modern one is in the awareness of the loss of references of the past, peculiar to modern man: "there's the whole difference between ancient tragedy and modern, Signor Meis – believe me – a hole torn in a paper sky" (Pirandello, 2004, p. 139), namely in the cognition of earthly reality and of no more possible footholds in unearthly dimensions. Nietzsche had also observed the correlation between the idea of tragedy and uncertainty, when he said that "even the clearest figure still retained a comet's tail which seemed to point to the uncertain, to darkness beyond illumination" (Nietzsche, 2000, p. 67).

The idea of this chapter is to resume the arguments mentioned in previous pages – of Modernity as a European crisis and revolution at all levels, which contributed to tragedy and uncertainty even in the geographic field – and extend them on a broader analysis, making some comparisons with literary characters that, according to two major authors such as Carl Schmitt and Agostino Lombardo, have been particularly meaningful for Modernity: Hamlet, Othello and Dr. Faustus. The aim is then to show the intrinsic problematic and uncertain nature of cartographic representation in Modernity, which resulted in what can be defined as a "cartographic tragedy" or, as others have identified it, as a "crisis of cartographic reason" (Farinelli, 2009).

It is possible to speak of a sort of tragedy of cartography derived from a deep contradiction, experienced by those who knew and represented the world, including its more accurate features, because in cartography – as in other sciences – a process of enhancement of increasingly scientific representative techniques began, but was not able to fill the void resulting from the loss of certainties of bygone references. That "certainty of representation" – as has already been defined in the arguments of Farinelli which refer to what is stated by Heidegger (Farinelli, 1992, p. 55) – that was acquired in Renaissance, paradoxically seems to be the representation of uncertainty deriving from the change of European perspective on the world, from that "loss of centre" which started with the discovery of America and the Age of Discovery.

The tragedy of Hamlet, Othello and Dr. Faustus (Schmitt, 2006; Watt, 1996), with the necessary adaptations, was *conceptually* the same as that of cartography: a reduction of knowledge of the world to intellectual, subjective and immanent levels, caused by the inability to read the profound geographic and political changes and a sort of *hybris* that marked the modern travellers.

If, at first glance, this parallelism may seem too daring, and maybe too "literary" for a geographic book, on a closer analysis, we will be able to observe their close proximity, from a symbolic point of view and of "images" representative of an era that would seem to complement each other well.

Maps in Modernity: instruments of uncertainty

In order to better understand the comparison proposed between emblematic characters of Modernity and modern cartography, it would be useful to retrace some essential stages in the evolution of cartography, from its main medieval features to modern ones, in order to advance useful comparisons and resume, even in more detail, that which has already been highlighted by Casti (1998) in reference to modern methods of geographic approach, "geographers in full sail" and "laboratory geography".

These two ways of coming to terms with the knowledge of the world characterized the approach of modern geographers, complementarily completing each other as both are necessary for coherent knowledge of the world. In the final comparison, these will be taken as symbols of the "modern cartographic tragedy", which was not only related to the purely representative level, but which had more general repercussions on political and social levels, because cartography, in Modernity, took on even more decisive roles and functions in contrast to the uncertain European political dynamics and those of discovery.[1]

Furthermore, the "cartographic tragedy" cited here was related both to loss of references of the past, that introduced a moment of loss of referential and symbolic certainties for man, and, speaking more generally, to the great explorations, to uncertainty of the boundaries of the new discovered lands and the indefiniteness that dominated the 16th and 17th centuries, concerning the boundaries of the world and the American shores (Buisseret, 2003), as seen in the previous chapter.

The map was in fact an indispensable instrument for travellers and brought with it all the uncertainty of "geographers in full sail". At the same time, the map was also used to escape from the uncertainty resulting from the exploratory travels and to confirm the discoveries made by travellers, to allow them to retrace sea and land routes, thus also contributing to the political logics underlying many of the modern travels (AA.VV., 1992).

In most cases, modern maps contained "uncertain features" of the lands represented (Lois, 2018), features that had to be redesigned and adjusted precisely outside traditional norms, as with the case of Columbus and the "accomplishment of the mad flight".[2] And indeed "the great geographical discoveries were made using maps that, showing uncertain features of undiscovered lands and unexplored straits, gave credit to a hypothesis that acquired credibility through the representation of those supposed geographical objects" (Casti, 1998, pp. 23–24).

From a perspective which is more related to the dynamics of power connected to the discoveries and in their cartographic representations, Jordan Branch stated that

> the geographic uncertainty of the discoveries, whether a new continent or a part of the known world, had to be circumvented, as the monarchs wished to assert their political claim in any case. Columbus' traditional means of asserting authority on the spot – as he declared, "by proclamation made and with the royal standard unfurled" – was an insufficient basis for claiming a poorly understood territory.
>
> (Branch, 2014, p. 106)

In the era of the great geographical discoveries, maps were also used to compensate the enormous uncertainty about the new lands, precisely from the geographical point of view:

> during the early modern period, many agreements on divisions of political claims were made in the face of great uncertainty or ambiguity regarding the actual divisions on the ground, a fact that made these resolutions easier, not harder, to achieve.
>
> (Ibid., p. 182)

In addition to the symbolic and representative factor peculiar of such times, the different roles of maps – political and as projecting tools – in different historical periods should be emphasized. Maps had an essential role, alongside the evolution of European political systems,[3] up to the development of the Nation State, above all from the medieval period to the modern one as already stated: "maps were a necessary – though not sufficient – condition to the emergence of the sovereign-state system" (Ibid., p. 5).

As noted by Claude Raffestin, taking possession of a space, concretely or abstractly (e.g. through representation), the actor "territorializes" the space (Raffestin, 1980): hence the strategic and essential importance of geographic knowledge in general, and in particular of cartography in Modernity, as a political instrument in the process of territoriality that was also related to the new discoveries: "there is no doubt that the cartographic document also covers a fundamental role in the use of the space, becoming an indispensable instrument of the 'discovery'" (Casti, 1998, p. 23).

The map, therefore, was also used to politically affirm State sovereignty on some "new territories" (Pagnini, 1985; Wood, 1992; Branch, 2014) and to try to delimit power, with relative certainty, within that which was still highly uncertain. Modern maps had, therefore, the function of offering a solution to political and geographical uncertainty which arose in the Modern Age, or at least to contain it. As stated by Raffestin, with the rise of the modern State, things change, also thanks to the appearance and vulgarization of an instrument of representation: the map. The map is the privileged instrument used to define, delimit and mark the border (Raffestin, 1980).

The central question in this chapter, and in the more general reflection proposed in this discussion, concerns the change of perspective and reference points passing from medieval maps to maps of the modern period, when the representative "certainty" seemed to fail to compensate for all the uncertainty resulting from the loss of symbolic elements of the Middle Ages.

Undoubtedly, medieval notions of space were characterized by imprecision and approximation. Moreover,

> [M]edieval man was disposed to see in numbers, not primary a means of reckoning, but rather a revelation of the divine harmony that ruled the world and hence a form of magic. So, medieval man's relationship with nature was not that of subject to object. Rather, it was a discovery of hims3elf in the external world, combined with a perception of the cosmos as subject. In the universe, man saw the same forces at work as he was aware of in himself. No clear boundaries separated man from the world: finding in the world an extension of himself, he discovers in himself an analogue of the universe. The one mirrors the other.
>
> (Gurevich, 1985, p. 56)

Reasoning according to Descartes' wax analogy, delving deeper in a conception which contemplates the repercussions that a religious and metaphysical vision had on men's lives alongside the perceivable elements, it seems that the symbolic references in medieval cartography – clear, sure, transcendental, even though not realistic in practical terms[4] (Harley & Woodward, 1987, p. XVI[5]; Scafi, 2006a) – do not set up a cartography capable of providing men with certainties, when as opposite the uncertainty regarding the New World's borders, lacking of any religious conception and yet so realistic, was really felt and could perfectly shape the modern mentality. Or, at least, medieval cartography was probably able to support a system of religious certainties on which man's whole life was centred. So, the representation was not based only on realistic elements but, as Jacques Le Goff noted, on that certainty which could be found in the imagination which maps and art representations furnished to men, even in abstract terms:

> la représentation est liée au processus d'abstraction. La représentation d'une cathédrale, c'est l'idée de cathédrale. L'imaginaire fait partie du champ de la représentation. Mais il y occupe la partie de la traduction non reproductrice, non simplement transposée en image de l'esprit, mais créatrice, poétique au sens étymologique.
>
> (Le Goff, 2013, p. 3)

Therefore, remaining with the Cartesian wax metaphor cited, the idea of uncertainty seems paradoxically more expressed in Modernity than in the Middle Ages, precisely because it is the representation of a system undergoing deep change and without a central axis, in which the discovery of a New World had permanently projected Europe outside itself and also as the result of a radical rethinking of the political, social and religious logics within Europe, in the political geography of the Old World that was assimilated with the ideas of *uncertainty* and *general crisis*.

There is also a strict and, in some way, paradoxical connection between abstraction and certainty of vision in the Middle Ages. This is remarked by Gurevich, when he states: "length, breadth and area were not determined by means of any absolute measures or standards in abstraction from the actual concrete situation" (Gurevich, 1985, p. 55). But it was synonymous of a solid and certain condition for man, as the Russian medievalist later pointed out:

[T]hese measures of area not only varied from one place to another; it never entered anyone's head to suppose that a relatively more accurate system of measuring land might be required. The accepted and universally widespread system of land measurement was completely satisfactory; it was the only possible, indeed only conceivable one, for the people of the Middle Ages.

(Ibid.)

This is a very clear statement about the certain ("the only possible, indeed only conceivable one") vision in medieval times, which was not certain from a concrete perspective, but *gave* people *certainties*.

The world in the image and likeness of God: the transcendence of the medieval cartographic vision

Before becoming an instrument of effective power and a *symbol* of modern uncertainty, maps lived a long period of full assimilation on a religious level, albeit not unrelated to medieval power dynamics. In this case, rather than dwell on the technical element of the different types of medieval maps, which have already been thoroughly and comprehensively addressed by Roberto Almagià, Aldo Sestini, John Brian Harley and many other scholars, or on the technical and historical evolution of maps in the Modern Age, it would be useful to focus on the metaphorical and symbolic level in the transition of the epoch, in order to find if, and to what extent, the uncertainty within Europe, related to the extension to the New World, was also transposed in the cartographic field, as already outlined in the previous chapters.

To do this, it will be necessary to depart from the simple objective element of adherence to reality, and try to read the map, its language and its symbolism in the light of a comprehensive vision, capable of contemplating the spirit and the mentality of the past epochs. I will therefore use some paradigmatic examples of the change of perspective that is already clear in existing literature, so it does not have to be traced in minute detail, but only in the general and the symbolic aspects considered most interesting here for the purposes of this dissertation.

The moment of transition from medieval to modern cartography can be conceptually represented as the transit from the *mappamundi* – symbolically understood as a medieval, representative modality which was highly and religiously metaphorical – to the *theatrum mundi* (see, e.g., Leti, 1690), the deeply realistic approach conforming to the "irreversible reality" of modern conception.

The idea of *mappamundi*, that is literally a "cloth" of the world, a canvas (Brotton, 2012) on which a representation of the globe was depicted (including its historical phases) and generally placed in manuscripts or in symbolic locations, seems to involve a certain idea of detachment from the world (Lilley, 2013), of mere "vision" of the same, filtered by a religious-metaphysical system (Harvey, 1991): as for the mapmakers,

> it seems that the medieval artist did not distinguish any too clearly between the earthly world and the supernatural world: both are represented with the same degree of clarity and precision in lively interaction within the confines of one and the same fresco or miniature. This is something very remote from what we call 'realism', Let us not forget, however, that the word "realism" is also of medieval provenance – but at that time the only "realities" were certain categories to which we today would deny reality.
>
> (Gurevich, 1985, pp. 6–7)

On the contrary, the concept of *theatrum mundi* fully implies the modern interpretation of reality, because the world was seen more as a stage than as a place of divine action, where, using the words of Shakespeare, "everyone is an actor and everyone plays a part", and where every man, therefore, takes an active part in the understanding and interpretation of the world (see Gillies, 1994).

In its various configurations of typology, the medieval map (Barber, 2006) had instead the main aim of providing an image of the world that was at the same time political,[6] religious, social and general,[7] corresponding to what European believers expected to receive and observe, also according to the Sacred Scriptures. As Gurevich stated, there is a strict connection between the Bible and cartographic representations during the Middle Ages. The fact that Jerusalem was placed at the centre of the maps has a clear root:

> [A]s it was written in the Old Testament book of the prophet Ezekiel (5:5): "Thus saith the Lord God, this is Jerusalem; I have set it in the midst of the nations and countries that are round about her". Accordingly, Jerusalem was moved in medieval maps to the centre of the Earth, and was accounted the "hub of the universe".
>
> (Gurevich, 1985, p. 73)

In other words, medieval representations (including maps) served essentially as an enforcement of medieval beliefs and of its religious structure of faith, and reflected a unique idea of the world (see Edson, 1997), in spatial vision throughout different historical periods.[8] They included essential biblical references and the main religious and symbolic cities and places, all steeped in metaphysical and metaphorical references, as in the pictorial representative field. Gurevich outlined this point of coincidence between the sacred representation and feelings of time and space in a unique vision, which could be substantiated as a "certain" one:

> [I]t is essential to remember that members of primitive society understood space and time not as a set of neutral coordinates, but rather as mysterious and powerful forces, governing all things – the lives of men, even the lives of the gods. Hence both space and time are axiologically and emotionally charged: time and space can be good or evil, beneficial for certain kinds of activity, dangerous or hostile to others; there is a sacral time, a time to make merry, a time for sacrifice, a time for the re-enactment of the myth connected with the return of "primordial" time; and in the same way there exists a sacral space, there are sacred places or whole worlds subject to special forces.
> (Gurevich, 1985, p. 29)

In his monumental work on the history of cartography, David Woodward states, in this regard, that "the primary purpose of these *mappaemundi*, as they are called, was to instruct the faithful about the significant events in Christian history rather than to record their precise locations" (Woodward, 1987b, p. 286). The same representative logic, both spatial and temporal, also characterized the theatre of that time for medieval dramatists: "if all times are possible, all spaces are possible: and the 'cycle' thus moves from original Chaos to Creation and the Garden, from Bethlehem to the Calvary, from Paradise to Hell" (Lombardo, 2005, p. 16).

Massimo Rossi goes as far as saying that medieval maps did not have the task of representing space, but only time: "in medieval representations there was not space, but time, there was a symbolic representation of reality, religious beliefs, routes, pilgrimages, things lasted" (Rossi, 2008, p. 15). In the statement "things lasted", there seems to be an implicit idea of basic certainty that I am are here attributing to the *mappaemundi*, a certainty originating from being able to impress Christian standpoints (spatial, figurative, symbolic and temporal) in the human mind with a single image, which could be defined as "all-encompassing".

The idea of certainty seems to be inserted very well in medieval representations and ways of life. Life in the Middle Ages was consecrated to certainty of a religious vision, as Gurevich perfectly stated:

> [T]his integrated nature of the medieval world-view, however, in no way guarantees its freedom from contradiction. The contrasts of eternal and temporal, sacred and sinful, soul and body, heavenly and earthly, which lay at the root of this world-view were also deeply imbedded in the social life of the period . . . The Christian worldview of the Middle Ages "transferred" the real contradictions to the higher plane of the all-embracing transcendental categories. On this plane, the contradictions were to be resolved in the fulfilment of earthly history, as a result of the Atonement, the return of the world, through its development in time, to the eternal. Hence theology gave the

90 Cartographic secularization

medieval community not only the highest generalisation but also its sanction, its justification and sanctification.

(Gurevich, 1985, p. 19)

Indeed, in accordance with Le Goff, that certainty could be found in every field of human action: "parchemin, encre, écriture, sceaux, etc., expriment plus qu'une représentation, ue imagination de la culture, de l'administration, du pouvoir" (Le Goff, 2013, p. 4).

A particularly revealing example of medieval mentality transposed on map is the Ebstorf *Mappamundi* (1284, Fig. 5.1), in which the figure of Christ dominates the entire map, with his face at the top, his hands on either side – left and right – and his feet at the bottom, almost forming a living "immense host" (Farinelli, 2009, p. 12). In this *mappamundi* (Harvey, 2006; Westrem, 2010; Kupfer, 2013), everything seems to refer to a guarantee of divine vision: the same certainty is defined

Figure 5.1 Ebstorf *Mappamundi* (1284)

by the presence of Christ and by his embrace of the world, and his own entirety is given by the establishment of a centre in Jerusalem (von den Brincken, 2006).

This is a crucial moment in a representative vision of the world that wishes to base itself and rely on unearthly certainties. The centre, Jerusalem, therefore, becomes symbolic and factual at the same time, for the lives of those who contemplate (and not just observe) that map, because everything originates from it and spreads: Anna-Dorothee von den Brincken wrote, to this purpose, that "Jerusalem belongs to the Christian history of salvation" (Ibid., p. 355) and, as stated by Rudolf Arnheim, who does not neglect cartographic reflections, "a central position conveys stability. At any other location, objects are possessed by vectors that point in one or another direction. The central object reposes in stillness even when within itself it expresses strong action" (Arnheim, 1982, p. 84).

Establishing a centre, both symbolic and real, religious and spiritual, means, therefore, to fix a definite element, that "reposes in stillness". It helps to establish and guarantee a certainty, especially if established on religious basis. When contemplating a map with an established centre in Jerusalem, the observer will support its structure of earthly and unearthly certainties, because, again quoting Arnheim, "elementary visual logic also dictates that the principal subject be placed in the middle [of the map, in our case]. There it sits clearly, securely, powerfully" (Ibid., p. 72). In this element of centrality of the holy city lies therefore, one of the certainties of medieval cartography, which will be lost in modern representations.

The positioning of a centre, both on the canvas and on a map, thus contributes to the clarity of the representation, to the certitude provided in the observer and to the power of that same representation, which will be bound to an apparatus of certainty strengthened by a metaphysical vision. At the same time, the centre provides a hierarchy of places, thus establishing primary importance in Jerusalem, with the rest of the world revolving around it, and not just symbolically. The intent is clearly to provide a guarantee of certainty and stability useful in the life of medieval man, contributing to world order and its image.

This order is also found in the absence of empty spaces in the Ebstorf *mappamundi*, and more generally in medieval cartography, to emphasize the impossibility of an awareness of emptiness, of any sort of absence therefore, of even the slightest form of uncertainty. If the centre provides certainty, the lack of empty spaces on the medieval map and on the Ebstorf map reconfirms this: in a world that is "in the image and likeness of God", voids are unthinkable and inconceivable. Le Goff well-underlined this aspect of medieval representation: "étudier l'imaginaire d'une société, c'est aller au fond de sa conscience et de son évolution historique. C'est aller à l'origine et à la nature profonde de l'homme, créé 'à l'image de Dieu'" (Le Goff, 2013, p. 6). That's why the Ebstorf map, like other medieval maps, does not present indefinite or empty spaces, or oceanic immensities: because they would represent an uncertainty that is not imaginable in a system which is the direct image of God, which is embraced by Christ, in the form of an "immense host": it would lose its most profound sense, based primarily on the certainty of that divine presence. However, if everything is emanated from the metaphysical and symbolic, religious and factual centre, the periphery cannot be vacuous, empty, therefore meaningless, but must necessarily be an emanation of the centre itself, and even if populated by monstrous figures, it cannot be left without elements. To this purpose, the concept of *horror vacui* should be translated and interpreted as the *refusal of emptiness* and not as the *terror of emptiness*, because emptiness could not be conceived.

Therefore, already in this first example, which is one of the most cited *mappamundi* of medieval production, we can find some key elements included within a context of representation that is intended to provide certainty to medieval man: certainty of divine presence in the known and unknown world through the figure of Christ in the four cardinal points, in the represented space-time religious dimension and in the organic unity of a world[9] that revolves around a fixed centre in Jerusalem and does not allow voids or fear of voids. In reference to the vision of time and space, united together (Edson, 1997), for man of the past "time has been 'spatialised', it is experienced in the same way as space; the present is not separated off from the main body of time, composed by the past and the future. Ancient man saw past and present stretching round him, in mutual penetration and clarification of each other" (Gurevich, 1985, p. 29).

In the case of the Hereford *mappamundi* (1276–1283, Fig. 5.2), we can find further and very clear proof of what has been stated earlier concerning the certainty of medieval cartography in the Ebstorf map. The idea of an afterlife is also present in the Hereford *mappamundi*[10] together with damnation and eternal blessing, through the souls placed on either side of the zenithal figure (Christ), who warn men about their eternal destiny: on the left, there are the damned, on the right, the righteous and below the Virgin Mary, who exhorts her son, saying:

> dear son, my bosom, from which you took on flesh and the breasts at which you sought the Virgin's milk, have mercy – as you yourself have pledged – on all those who have served me, since you made me the way of salvation.
> (Scafi, 2006b, p. 158; Simek, 1996; Westrem, 2010, p. 7)

The message of the map is in this case very strong: Mary turns to her son, showing him her womb where he was made flesh, almost in indirect reference to the world itself as an expression of God. We can also contemplate man's salvation and his damnation, and certainty is to be found precisely in this vision of osmosis of this world and other worlds and in the certainty of life after death.

The idea of death is not only present in the awareness a life after death, but also in the impossibility of conceiving of life outside the known world, which is the will of God: the letters M-O-R-S are placed on the circumference of the map itself, indicating that outside God's will only death is conceivable. In a world like that of Ebstorf, where a void is not contemplated and everything is ordered by the centre, where everything is in its right place according to the will of God and only death exists outside divine conception, there is no space for uncertainty, neither as a choice of God nor as a choice for man, as also the reference to the two supernatural worlds (heaven and hell, basically) wants to emphasize[11]: no chance is left to uncertainty.

Cartographic secularization 93

Figure 5.2 Hereford *Mappamundi* (1276–1283)

The figure of Christ himself does not coincide by chance – as it does in most of the medieval cartographic production known to us today – with the East that is where the sun rises, in a very clear reference to a divine correspondence of the world and to Christ with the solar element.

Such a representation is also the image of a world according to the will of God, as in the previous case: if in Ebstorf map, this idea was evocative because of the all-encompassing figure of Christ embracing the world, as an expression of divine will, in this one, the *mappa*, the cloth on which it was drawn up, was an ox skin where veins and traces of blood are almost visible signs of an "organic" world that seems to come to life.

In this further classic example of medieval mentality and cartographic representation, in the grains of a terrestrial organism, in the setting of a centre, in the presence of Christ and in certainty of an afterlife and that outside this world there can only be *mors* ("death"), the idea of certainty which dominated the life and mentality of men of that time are evident and very clear, with immediate elements that seem to fully confirm the idea of medieval representative stability.

The next passages of cartography would see an eruption of mundane and concrete elements, closer to reality, which would leave no space for medieval symbolic certainties. They would tear the veils of religion and supernatural and biblical reference, which had dominated the creative and expressive fields of medieval man, catapulting him into an "irreversible reality" that would open the doors to "modern tragedy". Man was, therefore, fully aware of the "hole torn in a paper sky", echoing the significant expression used by Pirandello, entering the domain of the modern world "tragedy".

The "middle-Earth": the *mappamundi* of Fra Mauro

Just a few years before the great geographical discoveries, in a passage considered emblematic in the history of cartography (except for the Portolan charts during the 14th century, which also represented a key moment in the representation of the world and in the development of its knowledge), Fra Mauro introduced substantial innovations, that are worth mentioning here in order to understand the passage from the certainty of the medieval *mappaemundi* to the uncertainty of modern cartography (Edson, 2007).

The *mappamundi* of Fra Mauro (1459, Fig. 5.3), because of its richness of symbols, constitutes a particularly illustrative point of intersection of the relative cartographies,[12] because it combines elements of the Middle Ages with typical elements of Modern Times,[13] in a confluence of mixed styles, references and innovations which is particularly exemplary:

> [T]he world map of Fra Mauro, is on the dividing line between the Middle Ages and Modernity and in this perspective, represents a very significant moment in the crisis of knowledge of the past and the gradual revelation of the new.
>
> (Falchetta, 2008, pp. 29–30)

The picture of the world as a perfect organism without empty spaces is still present, but at the same time, it has become more accurate in details and representation, because it attempts to adhere as closely as possible to reality – certainly not the priority in the Middle Ages. Even the choice of not including desert spaces is a choice, defined by Piero Falchetta, as "qualitative", in the importance given to represented space by the cartographer.

Figure 5.3 The *Mappamundi* of Fra Mauro (1459)

However, there are unreal figures which derive from a deeply rooted conception in religion at that historical moment. The map is orientated to the South, no longer to the East, as was almost always usual in the Middle Ages and has been seen in previous cases.

To support the view of those who consider this map an example of the transition to Modernity, the city of Jerusalem, a cornerstone of religions on Earth, above all of Christianity, loses its centrality (even if only slightly):

> the centre is marked by a metallic "umbilicus" and the convention of placing Jerusalem at the centre of the Earth as in medieval world maps appears to be outdated: here, as the author explains, the city "è i mezo de la terra abitabile secondo la latitudine dello abitabile benché secondo la longitudine ella sia più horientale. Ma perché la occidentale è più abitata se dici esser i mezo de li abitanti".
> (AA.VV., 1992, p. 176)

Furthermore, Fra Mauro no longer uses Latin, the universal language of the Christian Middle Ages, but Vulgar, the "particular" language (Falchetta, 2013; Zurla, 1806).

However, it is the Venetian monk's use of images of reality, from the experiences of travel and concrete elements, which brings this map closer to modern ones (Cowan, 1996). He declares this intention within the map itself, with explicit reference to *The travels of Marco Polo*, with the precise purpose of describing the world through realistic features, starting from experiential facts and not just from the metaphysical field, although it is still present,[14] going beyond the mental borders established in past epochs and strengthened by the mythical-religious medieval vision.

This is the case of the Pillars of Hercules: the cartographer slightly undermines the fact that they are insurmountable, almost instilling – without making mould-breaking claims – doubt in the kind of dogmatic approach of the past, anticipating, at least intellectually, the "accomplishment of the mad flight" discussed in the previous chapter, when he states that

> I have often heard many say that here there is a column with a hand and inscription that informs one that one cannot go beyond this point. But here I would like the Portuguese that sail this sea to say if what I have heard is true, because I am not so bold as to affirm it.[15]

It is also true that the reality of Fra Mauro is filtered by his belonging to the Catholic faith: he believes that his representation, although starting from realistic elements, will never be complete and systematic, because – he says – it is impossible to do that without having divine proof: "it is not possible for the human intellect, without the help of some higher demonstration".[16] In other words,

> [T]he human intellect, being limited, does not have such capabilities, and the thorough examination of this map, of its conformity to actual geographical reality is therefore, beyond his means; our desire for knowledge can only be tasted (*degustation*), it can certainly not achieve complete satisfaction.
>
> (Falchetta, 2008, p. 26)

The tone of Fra Mauro's hypothesis is clearly very cautious; he has not the audacity to suggest the crossing of that limit set by myth but de facto introduces this reflection, questioning those who sailed that sea, the Portuguese.

If the tones of Fra Mauro are very cautious and discreet, the symbols that he offers started a substantial revolution: evidence of this, that more than any other represents the transition of that era, is the shift of heaven from the top – where it was usually located, in coincidence with the East – to the lower left, in one of the outside corners of the globe.

Paradise can no longer find a place inside a world based on concrete factors and represented on the basis of travel experiences, but as a divine vision has to remain, so it is placed in a corner, in what Farinelli has well-described as the "expulsion from paradise", that perfectly conveys the idea put forward here: in a

> colourful map of the world that efficaciously marks the passage from medieval to the modern images of the world the passage that establishes the unprecedented novelty of his model consists precisely in this: for the first time heaven is placed in a corner, in the sense that while remaining in the drawing on the map, it is out of the circuit which marks the boundary of earthly images.
>
> (Farinelli, 2009, p. 23)

In that placing of paradise in a corner, we can find the essence of modern mentality that was surfacing in the European reality of the mid-15th-century, also marking the beginning of future geographical knowledge and of the explorations that would be intensified at the end of the century.

This change of perspective seems moreover to anticipate that uncertainty, including the cartographic one concerning the great voyages of exploration and their effects in the mindset of European man, geographic epistemology, conceptual categories, European centrality and, lastly, cartographic representation.

Cartographic secularization

All biblical references and divine elements that were predominant in medieval *mappaemundi* were abandoned during the Renaissance, from the horrific figures to the *horror vacui* (or *refusal of emptiness*), from the presence of Christ to spatial and temporal visions. There would no longer be a central, symbolic and metaphysical reference point in Jerusalem, nor would there be the historical information present in medieval cartography. All of the transcendence[17] elements were left behind, in order to make room for contingent and immanent realities, in a progressive shift of focus from divine things to terrestrial ones, with no more religious filters.

Man became aware of a new global model. In the world maps that were edited in the following stages, a realistic image of the known world was attempted and such representations had to be freed from the divine and transcendental images of the past: factual reality had to be represented as man knew it, without omnipresent supernatural references.

This process of reducing the representation to a realistic and immanent sphere, excluding every possible religious vision, led to a deep revolution in the European image of the world. The representation of this image of the world had undergone a radical change over a period of 50 years, from that transitional representation that was Fra Mauro's map of the world, that represented a moment of intersection between the two eras and that symbolically represented the peak and final stage of medieval cartography and a first manifestation of the modern one, to the representation of the beginning of the 16th-century.

In order to make way for realistic knowledge of the world, European man had to abandon the symbolic, factual and religious centre, embodied by Jerusalem and with that, every kind of religious certainty which used to rule man's lives. The certainty of a religious vision was replaced by concrete certainty, a realistic image of the world. Such a vision, devoid of any religious influx, later led to the definition of new a representative category which intended the world in an organic sense and was different to medieval organicity (both temporal and spatial): indeed "time usually doesn't appear on maps any more, and space finally masters the field" (Farinelli, 2009, p. 23).

According to this interpretation, space took on an exclusivity that it had not possessed before, becoming the most important element in the cartographic representative field. This image actually helps us to understand the idea of the full affirmation of geographical representation as opposed to the medieval "organic" one, in which space and time were used to share the limits imposed by geographical representations, in a univocal vision of human existence and life on Earth.

The passage put forth by Farinelli indeed identifies what could be defined as *cartographic secularization*, a process of progressive abandonment of religious features from the representation of the world, which coincided with the decades that marked the advent of Modernity.[18]

Cartographic secularization emphasizes the full acquisition of immanency in terrestrial representation, as opposed to the typical medieval vision. In Fra Mauro's *mappamundi*, the signs of affirmation of this secularization are recorded within a cartographic framework. This reflected the change that was occurring in every area of human existence and that found full statement during the 16th-century, also due to the revolutionary upheavals and the jolts from the Protestant Reformation.[19]

That moment, which was a transformation from a transcendental to an immanent vision, was crucial for Europeans and for knowledge of the world. It corresponded to the full development of factual reality and to its following affirmation, together with the extensive and dominant presence of the *saeculum* in contrast to the *sacrum*, of the *world* as opposed to the *otherworldly*, of the *time of history* as opposed to the *time of God*. In other words, an irremediable and distinct *caesura* of the two spheres occurred, and a univocal and organic feedback on the map would never again be found.

If *secularization* means that process through which man takes possession of his own means, detaching himself from God, in the individualization of existence and human knowledge, *cartographic secularization* refers to the same process of a centralisation of man as opposed to any transcendental vision in the representation of the world. Consequently, a cartographic self-centeredness was produced with time, as result of the individualistic setting of the image of the world, in which the centre is no longer the symbolic and absolute Jerusalem, but one established conventionally by men. The absolute reference to a symbolic place was lost, to make room for any possible solution, as inevitably happened in the course of the following centuries.

The truth, including the cartographic one, was unilaterally established by individual will and no longer by divine will or by the centre incarnated by the Church, because earthly things assumed absoluteness on the map, which did not have to contain divine messages for the whole of humanity, including warnings for mortal existence. The map was no longer a tool to establish universal messages, useful to man's orientation in life which contemplated both the human and the divine. It became a "simple" means of orientation in the world. It lost its spiritual function in order to assume a material one. *Cartographic secularization* was no longer a sign of an encyclopaedic and all-encompassing vision, but made use only of acquired knowledge and was based on human experience.

At the beginning, the model used in medieval conceptions that continuously swung between the certainties provided by dogmatic truths and the indefiniteness of cartographic representation was *abstraction*. Secularization imposed a new model which gave shape to the modern image of the world. Medieval maps were meant to guarantee and establish existential truths for humanity: they worked as a mediator between life on Earth and above the Earth, as there was no substantial difference between the profane world and the divine world, at least according to medieval conceptions, so much so that, in Ebstorf's Map, the world corresponded to the "mystical body of Christ". Not by chance Vladimiro Valerio underlines this aspect, when he affirms that "maps of the world of this kind correspond to the need to clarify and transmit through images, texts and religious content also to the illiterate, so they call on the evocative and mnemonic capacity of images" (Valerio, 2012, p. 217).

Indeed, cartographic literary sources were those related to sacred texts and the Bible in particular. Secularization could also be traced in the change of these sources, which definitively moved to the consultation of works written by travellers, to odeporic literature which represented the bases for modern geographers' and cartographers' *modus operandi*.

This moment of transition – clearly if not definitively represented by Fra Mauro's map – corresponded to a mooring of realistic representations of the Earth, in a more general systemic *crisis* framework than had previously occurred. It was determined by two events: the rediscovery of Ptolemy that changed the way of intending cartography and knowledge of the world and the discovery of America that transformed the shape of the Earth and European *forma mentis*.

These two events were the two pillars on which secularization was based.

The first happened at the beginning of the 15th-century and brought back the work of the Alexandrian geographer of the 2nd century to the European cultural world. The great change in the conception of cartography taken up by Ptolemy consisted in the use of geographical coordinates and scientific representative models and was propaedeutic to scientific cartography, so that a slow reconversion and revision of maps depending on the position of places took place, and the latitude measurements practices, which had never been fully abandoned during the Middle Ages, gained back importance as indicated by Ptolemy in his astronomic treaty. The data related to the increase of latitude determinations and cartographic projections, during the course of the 15th-century, were countless (Ibid., p. 224).

The restoration of geographical grid-references caused the representative models of the Middle Ages to progressively lose their spiritual references. They were now based on new human and scientific certainties, and metric certainties were established in order to portray a three-dimensional reality onto a flat surface. This was a cartographic dilemma that had not existed in the medieval world, because at that time people used to think four dimensionally as they had to take the supernatural dimension into account. Furthermore, during the Middle Ages, strictly representative dilemmas were not considered, as the only important issues were existential.

When this priority no longer existed, others arose, to almost compensate for an emptiness created by the secularization process. During the 15th and 26th centuries, with Ptolemy, a renewed representative asset, able to provide new certainties and giving great impetus to scientific geography and to the realistic conception of ecumenical reality, was established.

The other event that determined the final "crisis" of medieval cartography was the discovery of America, already addressed in terms of uncertainty in the previous chapter. This initiated a new era for European man and for the image of the world and it was the follow by a *forma mentis* that had already begun with the rediscovery of Ptolemy: in this sense, if the conviction of Columbus that he was acting on account of the divine – as was seen in Vespucci – was also a sign of his belonging to the medieval world, in the attraction he had towards the unknown, the clear mark of Modernity of which he would become the first symbol may be possibly foreseen.

According to this interpretation, the existing link between the revelation of Ptolemy's work and Columbus's adventure may also be evident, meaning that the result of a cultural background was useful and preparatory to the Age of Discovery.

These two events are the basis of the systemic change that occurred in modern geography and cartography, in the crisis of the previous medieval asset that reverberated in 15th-and 16th-century European culture and science. This new way of understanding the world, its laws and its representation, was founded on these two pillars and the *theatrum mundi* paradigm that replaced the *mappamundi*.

Notes

1. In Turco's opinion, maps do represent not only the orthodoxy but also a social need for cultural, economic and political reasons which they imply.
2. For an adequate knowledge of cartography in the aftermath of the discovery of the New World, see AA.VV., 1992.
3. "The epochal transformation of politics in early modern Europe – a shift from the complex authorities of the Middle Ages to the territorial exclusivity of the modern state – . . . suggests the importance of mapping to this process" (Branch, 2014, p. 5).
4. Edson called it "geographical inaccuracy" (Edson, 1997, p. 13).
5. They stated: "maps are graphic representation that facilitate a spatial understanding of things, concept, conditions, processes, or events in the human world" (Ibid.).
6. Here, the author underlines: "in the Middle Ages, however – unlike today – sacred and secular history were inseparable because Church and State were not divided but inextricably intermingled" (von den Brincken, 2006, p. 355).
7. Indeed: "Medieval images, which we would now refer to as maps, illustrated landscapes and presented information on matters such as administration and theology. These were not meant as generic sources of directions, but were intended for the needs of particular

individuals, institutions and occasions" (Ducza, 2013, p. 8). See also Barber & Harper, 2010; Edson, 1997; Harvey, 1991.
8 Because: "Christianity was a faith rooted in history, and the Bible was seen as the record of God's encounter with man, the imprint of the divine on the human world" (Withfield, 1994, p. 12).
9 The organic view of the relations between man and nature, man and the surrounding world, is well-expressed by Gurevich, when he writes: "this world picture was engendered by man's relationship with nature as an extension of his ego, and was very closely bound up also with the similar organic unity of the individual with the social group. It is an attitude which gradually evaporates *pari passu* with the transition to modern times" (Gurevich, 1985, p. 54).
10 See Birkholz, 2013; Harvey, 2002; 2006; Lilley, 2013; Reed Kline, 2001; Westrem, 2010.
11 See Scafi, 2006b; 2011; 2013.
12 Here, two main lines of interpretation diverge: those, as Leo Bagrow (1948), who think that the map represents the culmination of the cartographic interpretation – and therefore also of the worldview – of the Middle Ages and those who, like Franco Farinelli, for example, believe that it constitutes a substantial moment of transition between the medieval and the modern cartography.
13 In Woodward's opinion, the map "is far more secular in nature than, for example, the Ebstorf map. It is transitional in the sense that it included information derived from portolan charts, from Ptolemy's *Geography,* and from the new discoveries in Asia", even if he underlines that some authors (see Bagrow) pointed out that it seemed "the summit of Church cartography" (Woodward, 1987b, p. 315).
14 Fra Mauro will explain: "This work, created as an act of homage to this most illustrious Seignory, is not as complete as it should be because it is not possible for the human intellect, without the help of some higher demonstration, to verify completely this cosmography or that mappamundi; from these, one gets what is more like a sample of information rather than a full satisfaction of one's desire. So if someone contests the work because I have not followed Claudius Ptolemy either in the form of the world or in the measurement of latitude and longitude, I do not want to defend this map in any other way than that in which Ptolemy defends himself when, in the first chapter of the second book {of his Geography}, he says that one can only speak correctly of regions that are visited continually; of those which are less frequented no-one should think himself capable of speaking with equal accuracy. But if he meant by this that he had not been able to verify his cosmography thoroughly, because it is a long and difficult thing, and life is short and experience fallible, he is actually admitting that with the passage of time the work could be improved – that is, one could have more certain information than that available to him. So I say that in my own day I have been careful to verify the texts by practical experience, investigating for many years and frequenting persons worthy of faith, who have seen with their own eyes what I faithfully report above".
15 "Io ho più uolte aldido da/molti che qui è una colona/cu(m) una ma(n) che dimostra cu(m)/scriptum che de qui no(n) se uadi/più aua(n)ti. Ma qui uoglio che por/togalesi che nauegano/questo mar dicano se/l'è: uero quel che ho audi/to p(er)ché io no(n) ardisco affe(r)/marlo". For a more complete dissertation on the cartographic vision of the Atlantic between Middle Ages and Modernity, see Randles, in AA.VV., 1992.
16 "Certo non è possibile a l'intellecto human senza qualche superna demostration".
17 Indeed, as enlightened by Gurevich, regarding the distance from the perception of reality, also "the artists and poets of the Middle Ages ignore the visible facts of nature, they depict no landscapes, they pay no attention to individual peculiarities" (Gurevich, 1985, p. 5).
18 See Ricci, 2021.
19 See Ricci & Bilardi, 2020.

6 The tragedy of cartography in the Modern Age

From transcendence of *mappamundi* to immanence of *theatrum mundi*

We have already seen how the new world system was no longer symbolically centred on Jerusalem, around which the rest of the world revolved, but was based on an effort to provide a complete and truthful image of the world. At the end of the 16th-century, this process coincided with the first atlases (Goffart, 2003), created within the Dutch cartographic "*industria felix*" (Quaini, 2006, pp. 97–116) and developed between the 16th and 17th centuries in the Golden age of the United Provinces (see Schama, 1988[1]). These modern atlases sought to provide a different image of the world, no longer as a product of God, according to a metaphysical vision, but subject to study and contemplation of its realistic features, as seen in the best and most complete cartographic representations of the time (Goffart, 2003; Ricci, 2013b).

This kind of representation, which was named *Theatrum orbis terrarum* and that here is summarily included in the concept of *theatrum mundi*, was an attempt to systematize and organize the uncertainty that the great geographical discoveries had introduced in European knowledge of the world.[2]

But what does *theatrum mundi* mean, and why do its essential features differ so clearly from medieval cartography?

The logic of the *theatrum mundi*, which was greatly developed in many forms at that time, (mainly geographically, historically and of course theatrically) demanded radically different presuppositions from the medieval past, when the map acted almost as a *medium* for knowledge of the world understood in a supernatural way. The idea of *theatrum mundi* originated in the second half of the 16th-century, through the work of Abraham Ortelius (van den Broecke, 1996), who was the first cartographer of the time, shortly followed by Gerardus Mercator, to have drawn an atlas in 1570. His intent was to make world geographic representations available and accessible in simple, practical format using a single scale of reference (Goffart, 2003; Mangani, 2008).

Although in Modernity a reference to religion was still often present, the world was conceived as a theatre, in which man had an active role, both in knowledge and definition of its borders and in its cartographic representation. Carl Schmitt very significantly wrote that "according to the already strongly baroque sense of life at the time – towards 1600 – the whole world had become a stage, a theatrum mundi, theatrum naturae, theatrum europaeum, theatrum belli, theatrum fori" (Schmitt, 2006, p. 35).

The globe was no longer upheld by the mythical figure of Atlas, it was no longer an entity wanted by God in which every reference was considered certain as it was the result of divine will and every place was a religious symbol and embraced by Christ the Redeemer, but it was a world in which every certainty – the centrality of sacred places, divine will through the immanent action of Christ, the presence of an earthly paradise and even of monstrous figures – was lost.

There was therefore, space for vast extensions, for those "deserts" that had in no way been conceived before, which represented enormous oceanic spaces. These spaces no longer only surrounded the inhabited world[3]: in fact, "Modern Times, as already noted by Alexander von Humboldt, begin only when the horror of the void ceases in cartographers and white spots and incomplete profiles of unknown lands begin to appear on maps" (Quaini, 1992, p. 786).

The cartographic uncertainty experienced for most of the 16th-century seemed, therefore, to be associated with the discovery of new American lands and with knowledge of the world, with a new conception of it and, more generally, with the uncertainty experienced by European man in his own continental context, in terms of political and international relations, of those epistemological "storms" that geography was experiencing in terms of knowledge of the New World.

The Atlas of Ortelius presented an important new aspect, at least a symbolic one, on the title page[4]:

at the top of the title page of the first modern collection of maps, published in Rome by Antonio Lafreri between 1565 and 1570, Atlas was pictured, for the first time on the front cover of a geography volume, holding up a big globe, with one knee resting on the ground.

(Farinelli, 1992, p. 78)

However, in Mercator's *Atlas* (1595), the mythological character, "no longer occupies the top edge but stands at the centre of the front cover and is seated, he no longer sustains but measures a small globe with a compass which is resting on his leg" (Ibid.; see also Sestini, 1981).

This type of representation of Atlas, no longer intent on sustaining the world but measuring it, is highly significant in terms of the change of perspective adopted and fully internalized by Modernity which consisted in a scientifically valid representation of the world, with its measurement and subsequent depiction. The fact, then, that this "new Atlas" who represented a symbol of a coherent, homogeneous and comprehensive collection which was easily accessible to the reader and no longer relegated in large buildings or simply hung on walls, is a further step in the attempt to systematize the knowledge about the world and also, to some extent, to guarantee new fixed points within the dominant uncertainty that the discovery of the New World had introduced in Europe.

The same figure of Atlas measuring the world was supposed to represent a new model of certainty, just as it contributed to ensuring the rediscovery of Ptolemy and the mathematical and scientific applications of cartographic representation, in a global context that had lost all believable religious reference and which was on the verge of becoming a theatre and a stage of itself: according to Massimo Quaini:

> [I]t is clear that at the basis of all these manifestations of geography ("this wonderful profession"), there is the same project of unification of knowledge ("to put together . . . to measure and see, separately and all together") that moves every intellectual and their curiosity before the extraordinary opening of new spaces on Earth.
> (Quaini, 2006, p. 61)

What mostly interests us here about the concept of *theatrum mundi* and, therefore, the modern conception of Atlas,[5] can be traced in the attempt of Ortelius and Mercator to accomplish the task of making the knowledge of the world coherent and scientifically "certain" as much as possible, since it was under the true domain of uncertainty.

The world becomes a stage and tragedy bursts onto the scene

The world, according to scientists, geographers and intellectuals of the 16th- and 17th-centuries, needed to be represented as it was, without filters. Cosmographers had to portray an accurate picture of the world measured by Atlas, accessible to their contemporaries: the theatre

> was all the more a constituting part of contemporary reality, a piece of the present in a society which by and large perceived its actions as theatre, and made no special distinction between the immediacy of a performed play and its own lived actuality. Society itself was still sitting alongside on the raised platform. The play on the stage could easily appear as theatre within the theatre, a live play within the other, unmediated play of real life.
> (Schmitt, 2006, p. 36)

This process, which led man to consider himself an actor in a world undergoing profound change that was seen as a stage on which he played a major role, was immediately visible in the cartographic field, but also had active repercussions in literary and theatrical expressions, because the characteristics of Modernity were common to all areas: these were "dynamism, pluralism, conflict, immanence, versatility, criticism of the system and questioning of traditional principles and certainties" (Lombardo, 2001, p. 95).

Contextually, geographers wanted to portray a realistic and coherent picture of the world, with all the images that had been previously overshadowed by fantastic illustrations and mediated by divine vision. Also for Shakespeare,

[T]he imperfection of theatre is the imperfection of life, and therefore theatre identifies itself with life, it's not only its "mirror". The world is a stage, but the stage is the world . . . This identification of the stage with the world becomes the cornerstone of the following tragedies, firstly for Othello, as the cartographer evoked by John Donne in *A Valediction: of Weeping*: "On a round ball/A workman that hath copies by, can lay/An Europe, Afric, and an Asia,/And quickly make that, which was nothing, all".

(Lombardo, 2005, p. 86)

In the theatre of the world, as modern man conceived and understood the globe, the immanence of modern visions, with consequent loss of traditional certainties, had an enormous impact on the way man was conceived, on his position in the world (and therefore the same geography of the world) and on the meaning of life itself in *tragic* terms: in the great crisis that caused awareness of the end of every medieval certainty, the "irrevocable reality" emerged, seen as "the dumb rock against which the play breaks, and the surge of the truly tragic moves forward in a cloud of foam" (Schmitt, 2006, p. 39): this process of full and tragic emergence of irrevocable reality and of direct identification with its representation occurred in Shakespeare's theatre, but originated from the uncertainty experienced in the geographic field, then transposed onto cartography.

It started from modern uncertainty but was, at the same time, its source, in a mutual coincidence of tragic and uncertain elements that appeared more than ever as the paradigm of the Modern Age:

[T]he tragic only arises from an event which exists as an irreversible reality for all the participants: for the author, for the reciter and for the listener. A contrived fate is no fate. The most ingenious invention cannot change that. The core of the tragic events, the origin of the tragic authenticity is irrevocable in this respect: no mortal can imagine it, no genius can invent it.

(Ibid.)

And also, noting the different role played by man in the knowledge of the world, Schmitt argued that "the active human being of that era felt as if he were on a platform before an audience, and thought of himself and his activity along the stage dimensions of his performance . . . To act in public was to act in a theatre and consequently, it was a theatre performance" (Ibid., p. 35).

This transition – from the *transcendence* of the *mappamundi* to the *immanence* of the modern stage of *theatrum mundi*, connected to a reality of profound crisis and uncertainty – was affected by this irruption of "tragedy" of Modernity, which derived from the realization of the limitations of earthly things and of the modern human condition.

This tragic advent was also found in the cartographic field, because "if the world is a stage, it is also a book, and theatre becomes a dictionary and syllabary and a code book of the world, a great text that is at the same time subject and grounds of a great conflict" (Lombardo, 2001, p. 104). So, the theatre expressed this geopolitical "great conflict", which had its origin in the geographical dynamics of the Renaissance. That is why the parallelism proposed here seems to have a basis, under the same theme of the *crisis* of Modernity and the consequent *uncertainty* and *tragedy* experienced by modern man.

Uncertainty, transposed, became tragedy: firstly geographical, concerning the knowledge of the world, and then cartographic, of the representation of the globe. That *theatrum mundi*, in other words, was a representation dictated by tragic elements, which originated from the awareness of the limitations of the globe, of human life and condition, from the loss of transcendental certainties of divine and religious world order.

The close correlation between atlases, modern cartography, the idea of *theatrum mundi* and the work of Shakespeare is also underlined by John Gillies, who states that

> "Shakespeare's Globe" is interesting for another reason. A real theatre, it can tell us much about the conceptual character of the Ortelian *Theatrum*, and vice versa, for the reason that the theatrical metaphor is just as important in Ortelius as the cosmographic or "global" metaphor is in the discourse of the Elizabethan theatre.

According to the same author, "Shakespeare's theatre and Ortelian cosmography, I suggest, are dialogically related" (Gillies, 1994, p. 70).

Hamlet, or the tragedy of (laboratory) cartography and the loss of certainties

The transition from the medieval world of certainties to modern uncertainty is identifiable as a moment of tragic irruption of man into an immanent reality. The theatrical representation of this historic transition coincided with the conclusion of the "game", of what Carl Schmitt called the *trauer*, the game of fiction on a stage, without any possibility of distinction between representative plan and reality. This is what defines the tragic element: it "ends where play-acting begins, even if the play is meant to make us cry, a sorrowing play for a wretched audience, a deeply moving sad play" (Schmitt, 2006, p. 35).[6]

Shakespeare's Hamlet was conceived and written in a time of systemic change in England, in which

> America and other worlds are discovered, new science advances, while the feudal and aristocratic system collapses due to the drive of "new men". This is the reality to which Shakespeare belongs (and that he also helps create), this time of great crisis and transformation that represents the passage, in England, from the Middle Ages to the Modern Age and the Renaissance.
> (Lombardo, 2001, p. 95)

A time when, in Europe, the internal balance and relations with other countries changed: Hamlet, indeed, "is above all, part of a crucial moment in British political history, just before the death of Elizabeth (which would take place in 1603), the queen without heirs who had hitherto kept the balance between the various political, social, economic and religious forces". But

> "Hamlet, and Shakespeare's work in general, especially in his tragic period, are not only linked to this crisis. They are tied to a broader crisis and transformation, which in England, is the passage from the Middle Ages to the Modern Age – a crisis identified with the labour from which modern man is born", as well as in theatre, for the medieval dramatist, "if all times are possible, all spaces are possible: and the 'cycle' thus moves from the original Chaos to Creation and the Garden, from Bethlehem to the Calvary, from Paradise to Hell".
>
> (Lombardo, 2005, p. 41)

The same process of coincidence between reality and its representation, of loss of medieval harmony, of isolation in attempting to give a new meaning to the world, also occurred in the cartographic field, which is why the parallelism between the work of Shakespeare (and of other representative modern authors) and cartographic works appears almost immediate and meaningful.

In the *Presentation* to the Italian version of *Hamlet or Hecuba*, the book in which Carl Schmitt reads Shakespeare's *Hamlet* in a historical key and makes an analogy with James I Stuart, Carlo Galli states:

> [T]he great historical event in which Shakespeare found himself involved and which determined the fate of the Stuarts is therefore, this: the formation of the modern *nomos* of the Earth, namely the birth of the State through the wars of religion, and through the opposition of land and sea; in other words, the end of humanistic virtue, and in the end, also of the divine right of kings . . . This event is confusing: James lost an order, the traditional one, and did not understand that it was necessary to make a decision in order to create a new order.
>
> (Galli, 2012, p. 12)

According to Schmitt, who reads the work of Hamlet in a political key, "a king who by his character and destiny was himself the product of the dismemberment of his era was present in his concrete existence there, under the nose of the author of the tragedy" (Schmitt, 2006, p. 26).

The origin of Hamlet's disorientation (James I, for Schmitt) has a clear geographical reference, because there had been a "shifting of the pivot of the universe" (Galli, 2012, p. 13): it is the same discrepancy already cited when speaking about the European "loss of centre" and the consequent bewilderment of European man which had effects in the cartographic field in terms of uncertainty.

At that historical moment, according to Galli, the characterization is even more strongly geographic:

> "the modern world order, the modern *nomos* of the Earth is the balance between the British *Seenahme* (the 'conquest of the sea') and the *Landnahme* (the 'conquest of land') of the European States; this balance also produces – through war lines of friendship – a distinction between areas where interstate war is limited and conventional and non-European areas where war has no limits" and "its origin is an epochal catastrophe".
>
> (Ibid., p. 12)

This reflection opens up a further and purely geographical field of investigation, which will not be analysed here: the transition from a life on land to a life on the sea, which coincided exactly with the end of traditional order.

As pointed out by Galli, the tragic element, for Schmitt, coincides with immanence. This tragedy therefore occurs exactly when the medieval idea of space-time transcendence (then transposed onto cartography) was lost, in the transition from the certainties of the past to the uncertainties of the modern condition, which is reflected in political action and in geopolitical dynamics, as has been noted:

> [T]he tragedy that bursts into the game is the time of *nomos*, because the *nomos* contains in itself the original fracture, because the *nomos* is upheaval, re-orientation and not an extended, linear time-space (even if it is nowadays designed in this way) but rather tragic in its origin.
>
> (Ibid., p. 26)

It is no coincidence that Schmitt, as well as other great authors, focused his reflection on Hamlet, and it is no coincidence that Hamlet is for many a symbol of Modernity: he is the tragic hero par excellence, tormented by indecision (and therefore by uncertainty), who reduces his actions to immanent levels, thus entering an endemic and stagnant condition of indecision, because he had lost every certain religious points of reference:

> [I]n times of religious schisms, the world and its history lose their established forms, and a series of human problems becomes visible, on the basis of which no purely aesthetical consideration is any longer capable of producing the hero of a revenge drama.
>
> (Schmitt, 2006, p. 26)

Hamlet represents "the theatrical image of modern conscience at its birth, that on one hand, rejects the legacy of the past and on the other, tries to make its way through the labyrinth of the present. A present where every aspect of life (human and social relationships, affections and ideas, feelings and beliefs) is subjected to great doubt" (Lombardo, 2001, p. 97), which is the great modern uncertainty.

Universal history had lost its definite structures of the past and any kind of religious certainty, experiencing that "shifting of the universe's pivot" in the historical and geographical sense mentioned by Galli: in other words, "the historical reality is stronger than any aesthetics, and also stronger than the most original subject" (Schmitt, 2006, p. 26).

Uncertainty and doubt originated from the crisis of the system of the past, which was based on religious certainties and from the replacement of that mindset with a life based solely on "irrevocable reality", which did not guarantee the return of certainty, but, on the contrary, it further contributed to that tragic uncertainty.

The spirit of Shakespearean tragedy, and, more generally, the principles of modern man, can be found starting from the "shifting of the universe's pivot". It was a critical transition of identity, which led to a new and renewed conception of the tragic nature of modern action and thought, based primarily on the restlessness caused by the loss of medieval certainties and on the uncertainty of its new state, because "the man of the Reformation, the man of Michel de Montaigne and Giordano Bruno (very influential in England in those years) is alone with his reason and his conscience", in that disruptive process of individualization and relativism of knowledge, which led to a world that "man has to now face on his own, trying to figure out its meaning without the backing of universal harmony that had supported him in the Middle Ages and until a few decades before" (Lombardo, 2005, p. 42).

The doubt that plagues Hamlet is the doubt of he who has lost every certainty and no longer has any supports to lean on. He indeed wonders whether it is right to act and in what manner, and this is the doubt of modern man, who no longer bases his actions on the transcendental, religious and metaphysical vision, but on his own reason, on that Cartesian "cogito ergo sum" already cited: "the doubt is all the more distressing because they are not faced with a solid, recognizable world, but that floating, relative, elusive reality" (Ibid.). The modern reality, according to Lombardo, is fluctuating, fluid, thus uncertain, as it was defined in European political geography.

Hamlet, therefore, having to act in that reality – that is the same as experienced by Shakespeare in the transit from Middle Ages to Modernity – trying to discern on the basis of his own reason and not on divine reference, asks himself continuous questions substantiating doubt and uncertainty: these have dominated modern man since the loss of centrality due to the great geographical discoveries, and in England since the advent of "sea life" (see Rapp, 1975, p. 500).

Modern man

> confronts the mystery of reality and tries to penetrate it, and without the support of the certainties of the Middle Ages continually asks himself the meaning of things, of everything that surrounds him and the words that define it. Everyone, after all, not just Hamlet, here, asks questions; the drama is crowded with questions, nobody has certainties, and reality offers mystery and ambiguity to everybody.
>
> (Lombardo, 2005, p. 49)

These statements provide a perfect explanation of the spirit of the modern man of doubt, of continuous questions to substantiate the meaning of the world which has changed its points of reference. It is the attempt to find the meaning of the world and of human existence in the world in a general uncertainty which is the result of those changes, revolutions and crises deriving from the transition to Modernity. It is, in other words, the attempt to give shape to a *vacuum* of certainties created in the transition from one era to another which originated above all, geographically, and that came from the loss of a symbolic and concrete centre.

That is why the modern crises of representation, the tragedy fully expressed with Modernity coinciding with its representation of "irrevocable reality" and Hamlet's decisional crisis, are also found in the cartographic field: because man has lost the definite references of the past which gave meaning to his actions and the world itself, and now has to face a reality without the symbolic, metaphorical, representative concepts of past epochs. In the cartographic representation of the modern world, although "certain" and scientific, he left behind the certainty that he had in the past.

He attempted, it is true, to restore the centrality of Europe in the planispherical representations of the 16th-century, through the Mercator projection where the Old Continent was "oversized" and always at the centre of the world, but that was only an attempt to re-establish a certainty that was now inevitably and permanently lost. That attempt, in other words, actually meant the loss of centrality, by now completed, on which the complex system of medieval certainties and part of earthly existence were also based.

Therefore, man tried to compensate the lack of those certainties through the scientific nature of experiential proof, removing the supernatural and metaphysical and reducing knowledge to a "purely" intellectual level. It is what Hamlet did, essentially, and it is what also happened in the cartographic field.

The reduction of knowledge to the intellectual level, which carries with it the seed of endemic doubt, sometimes unsolvable, of deeper uncertainty and of tragedy, was also reflected in the cartographic field, where certainties and space, time and metaphysical cardinal points of the medieval vision were left out, and gave way to a more scientific representative point of view, but full of the uncertainties that dominated European man at the onset of Modernity.

As in the theatre, in Shakespeare's *Hamlet*, reality was staged in direct correspondence with world events and with the dynamics of man and his different personalities in Modern Times, so in cartography, global reality was represented in its nudity, in its factual concreteness, without the omnipresence of religious intermediaries, almost completely faithful to experiential facts and scientifically accurate in the contours of the globe. This coincidence with reality, however, substantiated modern tragedy that, in our field of interest, also involved the representative level of modern "secularized" cartography: exactly like Hamlet, it could not give answers, because "the world remains 'out of order', because modern man has no answers, no certainties" (Lombardo, 2005, p. 59). Therefore, modern cartographers could not fully comprehend the changes which were taking place, which derived from the great geographical discoveries, as they had lost the cognitive instruments and the certain references which had been useful in understanding the systemic revolution that was occurring in Europe.

They therefore tried to compensate this lack of certainty by finding new possible certainties: among these was the centrality of Europe on world maps of the 16th-century. This is only the most obvious sign: in the cartographic projection proposed by Mercator in the late 16th-century, the choice of the world barycentre, orientated to the north, fell back to its original context, namely the Dutch one: it was, of course, a strategic choice in favour of the rise of the Netherlands as a globally influential reality, but was conventionally adopted by other European cartographers.

The rediscovery of Ptolemy and the application of scientific methods to representation fulfilled the same need to compensate for loss of stability, and the use of geographical coordinates was a further attempt at establishing new, different certainties.

The story of Hamlet, in other words, seems to be the story of the "laboratory geographer", who reduces knowledge to a purely intellectual level. In the more general loss of modern certainties, he has the scientific tools to represent reality but not the metaphysical ones to explain it, interpret it, give it meaning and bring it back – as in the medieval past – to such certainties able to guide him in understanding the world. Hamlet, like the cartographer, knows reality, is fully aware of it, but asks himself continuous questions and has innumerable doubts, and so enters the domain of uncertainty and indecision, trying to make sense of it, to disentangle it: the answer will be intellectual, as it will for the modern cartographer, but both will be unable to provide definite answers, above all, to themselves.

The reason is that Hamlet no longer has points of reference to show him the way, and the cartographer no longer has interpretative instruments to make sense of the new global reality that he is experiencing, discovering and helping to divulge all over the world.

This same reduction will cause the tragedy of Hamlet's indecision and loss of certainties and references that could offer meaning to his contingent reality. For the cartographer, it will lead to what is definable as the cartographic modern tragedy, within the domain of uncertainty which includes the representation of the world.

Dr. Faust, or the state of uncertainty of the "geographers in full sail"

Contemporary to Shakespeare's *Hamlet* is the work of Christopher Marlowe, *Dr. Faust* (1590), another symbol of Modernity containing elements of extraordinary importance regarding the uncertainty of Modernity and a possible – not distant – parallelism with the geographic reality of the time.

Let us reflect, first of all, on one thing common to both works: they both look on the city of Wittenberg as the symbol of the European revolution and crisis of Modernity. It is where, on October 1517, Martin Luther hung 95 theses on the cathedral portal which started the Protestant Reformation, one of the causes of that same European political geography instability (Rabb, 1975; Nexon, 2009) discussed in previous chapters.

The modern tragedies mentioned here, metaphors of European Modernity and of the new human condition of individualism,[7] are more or less related to this

German city, symbol of internal crisis and struggles that hit Europe throughout the 16th-century, almost symbolizing a correspondence between the uncertainty that dominated Hamlet, Faustus' typically modern "intellectual heroicalness" and the situation in modern Europe, which was dominated by the religious and geographical internal crises and that of identity.

At the same time, however, Faust also represents the crisis of European man faced with the loss of centrality of religious elements which reduced his knowledge to the intellectual level ("Faust is very particularly the intellectual's myth", Watt, 1996, p. 40) willing to do anything ("Faust's individualism is merely a ceaseless and active search for experience, for the deed", Ibid., p. 206), even to give his own soul away, in the name of knowledge, of wisdom that he would otherwise be unable achieve:

> [G]oing beyond the intellectual instruments developed in the Middle Ages and the striving for new knowledge. The "break" lies in the fact that Marlowe places at the centre of the drama, not Christ, not a saint, not even a king or a prince whose "fall" is represented as *exemplum*, but a doctor from Lutheran Wittenberg, the city of the Reformation (and the city, of course, in which the young Hamlet studied), in other words an intellectual, whose "ruling passion" is precisely an intellectual passion, whose "action" is interior and whose sin is a sin of the mind.
>
> (Lombardo, 2005, pp. 20–21)

Dr. Faust attempts to fully understand the world, in total awareness of the finiteness of the human condition and without divine support to solve the questions and doubts – even of knowledge – of modern Europeans, anticipating Shakespeare in the consciousness of the tragedy of Modernity. Ian Watt well-studied the myth and figure of Faust, mostly focusing his attention on the emergence of individualism during the Modern Age, and understood a key point which is central in this dissertation, the desire for knowledge which dominated both Faustus and the modern geographers: "there is, perhaps, nothing more difficult in literature than to give reality to the three basic themes of the Faust myth – the excitement of knowledge, earthly beauty, and spiritual damnation" (Watt, 1996, p. 29).

In fact, "the modern man, the new man, is headed for the conquest of the world"[8] (Lombardo, 2005, p. 23), but such tension of knowledge will lead him to his death: in Marlowe, "the rise of Faust is represented, the achievement of his utopia, of his Promethean dream, but in fact the rise is a fall, the journey toward infinite, limitless knowledge and immortality is a journey toward death" (Ibid., p. 24). In practice, the Faustian desire for total knowledge of the world, in the absence of certain divine references, finds a solution in earthly experience, because "here, in this world which is now only earthly, immanent (despite the reference to symbols of the transcendent), death has no redemption, no consolation" (Ibid., p. 25). Furthermore, in a more geographical sense concerning the new discoveries, the yet unknown lands and the uncertainties arising from these, Lombardo points out that "behind Faust, there is the void, an infinite dark space of which Marlowe feels all the anguish, as does Hamlet when faced with the 'unexplored country'" (Ibid.).

Therefore, in Faust, we find awareness of the loss of every human certainty and the consequent attempt to overcome it through knowledge, which in his case can be reached through a magical rite, and go beyond the material aspect of his immanent condition with the pact with Mephistopheles. In the selling of his soul lies the tragic condition of modern man, especially clear in those dialogues where there is no solution to the immanent condition of man and where the same Mephistopheles warns that "all places shall be hell that is not heaven". Indeed,

> only magic remains. Magic has always stood as a promise to take the individual beyond the present limits of his knowledge. More particularly, it offers the very powers in which Faustus has found orthodox knowledge to be deficient: the philosopher's stone might give immortal life; and necromancy might raise the dead.
>
> (Watt, 1996, p. 34)

This appears to be the condition of those who, in Modern Times, were willing to face death for knowledge of the world and a return of its image which was as faithful as possible to reality. It is, in other words, the condition experienced by modern travellers, by "geographers in full sail" who having lost every certainty of transcendental vision that would give meaning to places and their lives were ready to leave all security (like Ulysses in his "mad flight") to face the open seas, to overcome the dogmas of the past and land on new territories, to know them and provide the cognitive instruments for cartographers to represent global reality in its realistic features. They got to know the world by treading its lands and facing the open seas, going beyond the dogmatic constraints of religious beliefs of the past.

These geographers, explorers and men of action had to refer to those who were able to represent this knowledge on paper, systematizing it, making use of representative techniques, comparing it to the knowledge and literature of the past and having thorough familiarity with the sources. The geographer in full sail necessarily had to converse, discuss and engage with the one in the "laboratory", with those who knew how to intellectually maximize knowledge derived from the voyages of discovery, and most importantly, to translate it onto maps, thus contributing to the knowledge and the representation of the image of the world.

Without the certainty of metaphorical representation and the centrality of some places due to the awareness of the globality of the world and of the subjectivity and "geographic self-centredness", certainty had to be found again on an immanent, concrete and scientific level. This is what Faust did, in his enthusiasm for knowledge, to fill the gap of uncertainty deriving from the limitations of his earthly life (although, as noted, there are also supernatural references), and this is what the great explorers of the past did, who also – as has been previously pointed out – provided some religious connections to their own actions.

Similarly to the doctor from Wittenberg, they were willing to face death and the hazards of the travels into the unknown, to contribute to the knowledge of the world in the general loss of certainties that occurred in Modernity, in an attempt to provide a new and renewed meaning of the world and of worldly things. Both of these figures – Faust and the "geographer in full sail" – acted to provide answers to the doubts and uncertainties caused by the new knowledge of the world in the Modernity of great geographical discoveries. They tried to give certainties that had been irretrievably lost in the immanent view of the world.

Both tried to interpret the changes occurring at that moment in history, harbingers of uncertainty and doubts that Shakespeare's Hamlet would experience.

Othello, or rather the inability to read the world

Both in the Shakespearian *Hamlet* and in Marlowe's *Dr. Faust*, there are elements of the tragedy of man in the modern, scientific revolutions and crises of identity caused also by the geographical discoveries, which are easily transposable to other levels of human knowledge, such as cartography and geography in general, at least symbolically and metaphorically.

Hamlet, in his indecision to take action, in the uncertainty of his knowledge and with the lack of reference points to discern the situation in which he found himself, was compared to the intellectual level of human knowledge represented by the "laboratory geographers", who tried to figuratively translate the knowledge gained from European travellers; while, Faust was instead interpreted in the perspective of modern travellers in full sail who, willing to do anything – even risking eternal damnation – for knowledge of the world, left everything in order to make up for the lack of certainty and reach a global knowledge of earthly things, to finally be able to make sense of them, just as many of the travellers in full sail did during Modernity.

Another Shakespearean character embodies the problematic nature of Modernity that new geographical knowledge had brought with it in the 16th-century, including uncertainty in a new interpretation of changes that were affecting Europe and the world. Othello represents the tragedy of those who, faced with the revolutionary events of the Modern Age and having to deal with structural changes which were taking place, are totally unable to read and interpret them correctly, as he is still tied to the dynamics and mental structures of the past, and so sink into the final tragedy.

The tragedy of Othello lies in his inability to know how to interpret the deceptive language of Iago, who manipulates reality to his advantage through language, in the subjectivity of interpretation that is a cornerstone of Modernity and, at the same time, also of cartographic expression. Othello "is victim of the infernal and 'theatrical' intelligence of Iago, of his diabolical ability to use words (we could say, words of the theatre) to deceive – not only Othello, but all those around him, from Roderigo to Emilia" (Lombardo, 2005, p. 65).

The Moor of Venice "becomes the victim of Iago because he is too credulous, he is blind to reality, he does not know how to read the world. He is immersed in nostalgia and has remote memories of an almost fabulous past of glory and majesty and magic" (Ibid.). He is anchored to an old world of certainties and chivalric beliefs, belonging to a past that is now behind him, and the tragedy occurs because of his inability to read the changes.

Othello represents those who cling to a system of certainties that no longer existed in Modernity. They are totally anachronistic – even in their rightness, honesty and rectitude – and inevitably sink to final tragedy, which is a tragedy resulting from a crisis of language. It is the same crisis that Don Quixote experiences when he tries to follow a model of the past that makes him seem ridiculous in the eyes of his contemporaries, a system of literary and chivalric cardinal points which are essential for him to read and interpret the world: and indeed "there was no other more certain story in the world for him" (Ibid., p. 67).

Othello, like Don Quixote, is still tied to the certainties of the past, he is so incapable of interpreting Iago's deceptive language and reading the changes of the world around him that he enters into an almost parallel, false and unrealistic dimension: "the world collapses around him. Reality disappears. Before him there is a non-existent reality that he believes true, and that also destroys the reality of the dream, of the story in which he thought to be" (Ibid., p. 70). It is a crisis of language that leads to the collapse of the world surrounding him, to a global crisis resulting from his own anachronism and naivety when faced with a proposition of a distorted, individualistic and relative message, which was no longer universal and certain. This is the crisis and consequent tragedy of Othello, who seems to be the literary representation of the cartographic crisis experienced in Modernity after the loss of the certain medieval references.

The subjectivity brought about by Iago is the same relative vision experienced by cartographers in the Modern Age after the discovery of the New World and of previously unknown realities, which led to the cognition of the globality of the world and of a geographical relativism unable to give definite answers to the doubts of modern man and to his loss of references. This is the same crisis that would lead to the cartographic *critical* interpretation, the meaning of which is summed up in this statement: "what we read on a map is as much related to an invisible social world and to ideology as it is to phenomena seen and measured in the landscape" (Harley, 2001, pp. 35–36).

It is, first of all, a geographic tragedy, as rightly recalled by Lombardo, because "blindness" is the cause of the tragedy, and of the madness, and of the death of Othello. A blindness, a loss of contact with reality that reminds us of the darkening of the world described by John Donne in *An anatomy of the World*: "And freely men confess that this world's spent,/When in the planets and the firmament/They seek so many new; they see that this/Is crumbled out again to his atomies./'Tis all in pieces, all coherence gone,/All just supply, and all relation".

The great geographical discoveries, together with philosophical and scientific upheavals, had radically transformed the way modern man understood the world, and those who were unable to read these changes were inevitably subject to the final tragedy. The changes occurring at that time of passage demanded the achievement of a New World order and new logics and languages that regulated it: subjective, relative and therefore, also uncertain, because no longer universal.

The language of Iago, misleading, subjective and entirely relative, translated on a different level, is the same language that cartographers would use, also for power purposes. It was a representation that was no longer dictated by objectivity or by the certainty of religious references, by the univocal centrality of Jerusalem, but by a strong "egocentricity", just as philosophy and the cognitive approach. This was also due to the direct experience of "cogito ergo sum" repeatedly cited here.

In fact,

> it is precisely because the blindness, the bewilderment and the confusion of Othello are not only character traits but also wonderfully incisive scenic signs of a more universal precariousness [read also *uncertainty*], of difficulty in reading the cosmos that is characteristic of this period of transformation.
> (Lombardo, 2005, p. 75)

This statement by Lombardo contains the symbol of this tragic passage towards uncertainty, which is the same uncertainty found in the cartographic and geographical reading of the world.

It is, in other words, what happens in the cited passage of John Donne, which emphasizes the subjectivity of the cartographic vision, of his language, like that of Othello, thanks to which it is possible to "make that, which was nothing, *all*", maximizing and sublimating the uncertain, subjective and relative logic of the cartographic message, because the world in which they lived was like the world of Shakespeare, "who does not have an Olympus above him, has no certainties or fixed points of reference" (Ibid., p. 88).

Theatres of the world, cartography, tragedy and Modernity

The statements proposed here are theoretical considerations about a context that experienced a period of radical crisis and rethinking of the interpretative and representative models of the world.

In order to keep these considerations on a theoretical level, I have avoided making direct references to specific examples and tried to provide the conceptual instruments needed to understand the deeper, general and universal dynamism experienced in the transition from the Middle Ages to Modernity.

This critical and revolutionary moment has been likened to the transition of two cartographic categories which were typical of the first and of the second epoch: the *mappamundi*, which was an idea of the world and its symbolic representation derived from a religious medieval interpretation, and the *theatrum mundi*, which was the image of a globe that was gradually becoming known and portrayed in its realistic, coherent and scientific features.

I have discussed only the main innumerable changes which occurred in the transition between the two conceptions, which are significant of the shift of horizon experienced with the advent of Modernity.

The passage from the *transcendental* vision of the Middle Ages (which looked to the divine) to the *immanent* vision, horizontal and geared towards the things of the Earth (which characterized Modernity), caused the consequent representation of the world to undergo a profound rethinking that took shape from reference models which were no longer certain, but referred to a more general context of modern uncertainty.

In the medieval world, Christ was a living presence in the church. There was no vacuum, no error, nothing lacking, yet evil or diversity was present in the guise of monstrous, mythical or imaginary figures: that world could not conceive the idea of uncertainty (understood as void), although from a technical, representative and merely cartographic standpoint which can be seen today, medieval cartography presented many "uncertainties" and "inaccuracies",[9] sometimes even in the dislocation of continents.

In a conception of the world which was no longer the place for merely divine action but had become a *theatrum* in which everyone played an active role even in the field of knowledge, the attempt of the Modern Age to systematize knowledge and provide a realistic and coherent image of the known globe with the work of Ortelius and Mercator can be read as an attempt to compensate for a loss of transcendental certainties with scientific, realistic and representative references (Goffart, 2003).

But in the same *theatrum* vision of the world, which is found in many of the geographical and historical-geographical works of the Renaissance, there is also an additional element of reflection: if the world is a stage, the same actors that had been the symbols of those changes and revolutions of Modernity may represent the changes that intervened in the fields of cartography and geographic knowledge.

Here, therefore, lies the meaning of the parallel between Hamlet, the hero who asks himself "the great questions about the meaning of man and God, of life and death, of the word itself" (Lombardo, 2001, p. 97), and the cartographers, who reduced knowledge to an intellectual level, closed inside their laboratories, in which they developed their own representations of the world on the basis of travel documents and diaries provided by the great travellers, asking questions and seeking answers about the meaning of the world and the meanings that they could offer the readers who were lost without the references of the past.

The analogy seems even more pertinent if we reflect on the semiotic categories used in both fields: Hamlet's conflict is "the conflict between 'reality' and 'appearance', between 'to be' and 'to appear'" and "this existential drama that is at the heart of all Shakespearean tragedies is expressed above all, in terms of a quest for linguistic drama" (Ibid., p. 103).

It is a choice – a linguistic, semiotic choice, of the signs to put on the map – the cartographers also had to make in Modern Times, as they had lost the language of the past, the medieval signs and the religious symbols which had kept them in a world of certainties. Some residue of symbolization would also appear in the cartographic Renaissance (Van Duzer, 2013), but it would be a mere residue which would be gradually lost over time in the secularization process.

Cartographic language would experience the same crisis Hamlet experienced: "the journey of Hamlet in search of the lost meaning of the world is first of all a journey through words" (Lombardo, 2001, p. 103). The same thing would happen in cartography: it would indeed attempt to give a new, certain and clear meaning to the world, on the basis of a new – and increasingly scientific – semiotic system. Hamlet's search for certainties and answers is the same search for certainties and a new interpretative language which took place in the cartographic field of Modernity.

Language and its use is also at the heart of Othello's tragedy, which, in this sense, appears to be closely related to Hamlet's. Othello ends in tragedy because he has lost the explanatory instruments to read and make sense of the changes occurring around him. His tragedy originates from Iago's deceptive language. Without the certainties and the universality of medieval cartographic language, modern man has to try to give meaning to the changes of the world through new languages and renewed possibilities of interpretation. Furthermore, the comparison between Othello and cartography is related to a crisis that is embodied in the subjectivation of cartographic interpretation and representation, as well as the subjective – and misleading – language used by Iago.

His use of linguistic nuances to his liking and for his own purposes is the same as, in the long term, the crisis of cartography, when universalistic and objectively "certain" medieval references were abandoned to embrace a subjective model (the example of the Mercator projection with Europe at the centre is paradigmatic), which led to the rise of multiple representative canons, signs of different interpretations of world representation and their meanings, even in a political sense (see Boria, 2007; 2012).

The crisis of modern language was also a crisis of interpretation: as rightly noted by Quaini, the "shift of view" characterized the transition from the Middle Ages to Modernity, where the interpretation of an image was no longer purely visual but was tied to the evocative and imaginative power of words:

> [M]edieval man believes that imagination and beauty are blocked by the borders of places set by sight and drawing and that it is necessary to evoke the fabulous power of the spoken and heard, word. It was no coincidence that modern cartographic history coincides with the history of the emancipation of drawings based on words.
>
> (Quaini, 1992, p. 786)

Patrick Gautier Dalché stated that medieval *mappaemundi*

> represent not only historical events but also phenomena or being of the natural world and they could not have been understood without the help of accompanying texts. Nevertheless, the non-geographical contents of the maps are always written within a spatial frameworks.
>
> (Dalché, 2006, p. 224)[10]

This leads us once again to believe that modern uncertainty, which also dominates Othello's tragic context, is also present in cartography, its language and relative interpretation.

Finally, the tragedy of Dr. Faust – which not by chance takes place in Wittenberg, the city where the Reformation and modern religious crisis originated – seems related to the context of "geographers in full sail", who were willing to do anything to acquire knowledge and were able to face death and the risks of the voyages of discovery completely within the domain of the uncertainty of the journey.

This confusion of levels – theatrical and literary on the one hand and cartographic on the other – would seem to effectively describe uncertainty and the resulting tragedy that struck European man in his most deep-rooted and profound convictions, and that also occurred in the field of knowledge of the world and of its cartographic representation.

A tragedy was happening in the European *theatrum mundi*, in which man took an active part in its knowledge. It was due to the opening up of European spaces, to the change of geographical perspective, to the adoption of different, multiple and subjective standards of interpretation of the world and its configuration and to the lack of a definite centre: these were the same things that brought Don Quixote to live his life anachronistically as a knight-errant, and that led him to fight against windmills in a timeless battle.

Modern tragedy that marked all the characters mentioned – tragedy of language, knowledge, science and doubt, of uncertainty and of attempts to provide certain answers when all certainty was lost – turned into a cartographic representative tragedy: firstly because it was impossible to define the New World with certainty, secondly due to the positioning of voids in correspondence to unknown lands and the indefiniteness of American borders, and, finally, in the progressive "subjectivism" of cartographic interpretation.

Uncertainty becomes tragedy and tragedy is based on uncertainty, on the instability of a world that had changed horizons, standards of interpretation and centres of reference: this happened in the Modernity of Shakespeare, Marlowe and Miguel de Cervantes. It also happened – tragically and relentlessly under an aura of uncertainty – even in modern cartographic representation.

Notes

1 For the cartographic aspects, see also Ricci, 2013b.
2 The latter, incidentally, took the form not only in the loss of previous certainties, but also in the veracity of some historical maps, as the Vinland Map, which – if it had been believed to be true – would have undermined the whole generally known system about the "discovery" of America by Columbus. Ilaria Luzzana Caraci, among others, has focused on the issue, rightly stating that "cartography is a too tasty dish for those wishing to promote new theories about the discovery of America and more generally about the great geographical discoveries" (Caraci, 2007, p. 89). As stated by Gaetano Ferro: "[T]here are plenty of sources and testimonies: some – and not in small number – dating from the same Colombo..., but few are original and safe, because interpolations, manipulations, distortions and additions are not lacking and lead to doubt, suspicion and uncertainty at every turn" (Ferro, 1992, p. 381).
3 See, as an example at this regard, the *Carta da navigar per le isole Novamente trovate in la parte de l'India (Carta Cantino)* of 1501–1502, in which the vastness of the ocean space seems only to make possible to imagine the tremendous revolution in the vision of

the world lived in that historic moment: comparing this map with the Fra Mauro map, it is possible to notice the amazing perspective difference occurred to the European man in less than 50 years. The Anonymous *Planisphere* of the first half of the 16th-century (after 1522), appears extraordinarily indicative of the extreme cartographic "uncertainty" experienced in the Modern Age, found in the lack of definition of the entire western part of the map, which is dominated by the emptiness on the western lands to that of the New World (see AA.VV., 1992, pp. 716–717).

4   The writing stated: "the horse is born to tow and carry burdens, the ox to work with the plow and the yoke, the dog for hunting and guarding the houses, but man was born to study with his intellect the system of the universe and to meditate on his creation" (see Quaini, 2006, p. 71), as an evidence of the persistence of some religious sensibility, although understood in terms radically different from the past.

5   The same concept of atlas will be developed by Mercator, who will call atlas his complete and organic collection of maps of the world, which will be published in 1585. In contrast with that of Ortelius, the collation of Mercator is characterized by the originality of the maps directly depicted by him (see Ibid., p. 69).

6   About the concept of *trauerspiel*, it is not possible to forget the Benjamin's work (1977), in which he focused his attention in chapter II.

7   In Watt's opinion: "Marlowe's Faustus can reasonably stand as a prototype of one of the central, and rather unreal, assumptions of modern secular individualism. This individualism has retained some of the transcendental aspirations of both the Renaissance and the Reformation, but it cannot find or create a world in which they can be carried out; only magic can do that" (Watt, 1996, p. 40).

8   The same Agostino Lombardo underlines very significant passages at this regard: "How am I glutted with conceit of this!/Shall I make spirits fetch me what I please,/Resolve me of all ambiguities, /Perform what desperate enterprise I will?", Act 1, Scene 1, Lines 77–80.

9   Even if, as it is clear and as stated by John K. Wright, also mentioned in the work of Harley, the accuracy was not that the primary intent of those who drew up papers in the Middle Ages (see Wright, 1925).

10 Concerning the relationship between cartography and graphic representation, Jerry Brotton has noted that in the several considered languages, the word "map" is always linked to a visual expression (Brotton, 2012). He traced the languages lines of the word "map", to a direct image in front of the ones who look at it. In the immediate depiction of the terrestrial globe, in fact, the cartographer can give an image of the world in a very direct way. That image comes, above all, from a vision of it in a precise historical moment. The ones who see the map are led to read it from the cartographer's perspective that corresponds always to a message, which can be cultural, political, functional.

7 Uncertainty as a paradigm of Modern Times

Does a geography of uncertainty really exist?

At this point of the discussion, at the end of our diachronic and theoretical itinerary, we will put forward some conclusions and summarize the discourse. We started from a current reflection: uncertainty today seems to be a key-concept in reading contemporary social and political world reality. Many disciplines often associate it to *chaos* and *disorder*, *indefiniteness*, *fluidity* and *crisis* and used to describe the modern financial-economic reality and international relations in post-modern Western society.

The question posed is: how much uncertainty is there in Modernity and, at the same time, how much of that uncertainty originated in the geographical field?

Geography, more specifically political and economic geography – in a global analysis but also in its various regional determinations – seems to be directly connected to this speculation, perhaps even more so than studies have shown, if an interpretative key of uncertainty is applied. That is why it has seemed particularly useful to focus on the origins and configurations of modern geography of uncertainty and on its possible developments or current evolutions.

Modern uncertainties

According to the analysis reported on the first pages of this discussion, two distinctive and parallel moments, connected to each other, of the first general configuration have been identified.

On the one hand, the discovery of America, as an incipient moment of profound change in European perspectives (Galli, 2001), with the "decentralization" of the Old Continent and the consequent shift of the European centre of gravity[1] which corresponded to a sort of geographic and cartographic "identity crisis". On the other hand, the political, social and geographic-political (one might say *geopolitical*) internal upset of Europe in the Early Modern period, which, according to the interpretation given by Schmitt, originated from the spatial revolution of Modernity.

With regard to the first configuration of geography of uncertainty, corresponding to the radical change in the image of the world, the feature of indefiniteness is

DOI: 10.4324/9781003394204-7

found in the geographic "disorientation" of perspective and consciousness of the world, with its strong Eurocentric conception centred on the Mediterranean that had dictated the cultural guidelines in previous eras. For centuries, the continent had been centred on the Mediterranean which saw itself as the centre of religious dynamics compared to the rest of the world and at the same time identified itself as the West. Indeed, for centuries, the Mediterranean was the pivot area in the European mindset.[2]

This system of beliefs and certainties (Le Roy Ladurie, 1980) had to give way to an awareness of the existence of other world areas, suddenly geographically reshaping Europe within world dynamics and features, included simply in territorial quantitative terms, although some consideration of European centrality remained, especially in the cartographic field. All this had its repercussions not only in the way of understanding and studying the world (therefore concerning the geographical discipline and its categories) but also in the way the globe was represented: in this regard, what has been defined here as a kind of "cartography of uncertainty" is to be intended, in a comparison with the rise of cartographic secularization, as the rise of tragic element that marked the modern revolution.[3]

After the acquisition of a new, full awareness of the "new" size of the Earth and its globality, the geographic patterns of reference used to read and understand the world, underwent a profound process of change, which corresponded, in a representative aspect, to the affirmation of the *tragic* element: namely, the loss of certainties of the past. It was the geographic and cartographic transposition of what happened in literary suggestions, in theatrical and more generally artistic and cultural expressions that was the most direct symptom of the tragic condition of deep *crisis* that marked Modernity.

In the second configuration of geography of uncertainty, concerning the geopolitical internal dimension of the Old Continent, the European system which emerged from the revolution of the *global lines* required a thorough review, which was turbulent, often violent, and was shaped in terms of a redefinition of European inter-State boundaries, also due to those cultural upheavals of systemic interpretation started by the great oceanic expeditions and which saw their epilogue, *de facto*, with the Westphalia Treaty of 1648, because:

> no sooner had the contours of the Earth emerged as a real globe – not just sensed as myth, but apprehensible as fact and measurable as space – than there arose a wholly new and hitherto unimaginable problem: the spatial ordering of the Earth in terms of international law. The new global image, resulting from the circumnavigation of the Earth and the great discoveries of the 15th and 16th centuries, required a new spatial order.
> (Schmitt, 2003, p. 86)

According to Schmitt, the European modern crisis derived from a new political system of references, founded on new certainties, the ones of the *Jus Publicum*

Europaeum: State, territory, borders, common identity, nationality, common language and culture, etc.

As a consequence, the internal reorganization which culminated with the Westphalian Treaty, was an attempt to contain and reduce the uncertainty caused by the crisis due to the conflict and to the presence of multiple States (Tilly, 1994) which did not constitute a unitary and ordered system (Arrighi, 1994), but a substantial anarchical one (Bull, 2002). The political plurality of modern Europe had also appeared – in different shapes – in previous epochs, not least in the European Middle Ages. As already clarified, the multiple presence of political entities in the European political landscape of the Middle Ages would not seem attributable to a situation of *systemic chaos* as it was in the modern reality of the 16th and 17th centuries, but to a comprehensive set of laws, customs and religion that constituted a reality tending to unity and universalism (Gurevich, 1985; Le Goff, 2013). Indeed, the political structures that emerged during Modernity originated from the decadence and disintegration of the medieval system of laws, as well-clarified by Arrighi (1994, p. 31) and Brunner (1980, p. 38).[4]

The upheaval of cultural categories in the transition from the medieval to the modern world caused the disruption of the previous European geopolitical system, corresponding to the rise of State territoriality as a new political configuration of Modernity, opposed to the preceding one.

The *general crisis* or *systemic chaos* happened because of the supplanting of an *old* system which did not find new models able to guarantee a *new* order, in which "a new set of rules and norms of behaviour is imposed on, or grows from within, an older set of rules and norms without displacing it, or because of a combination of these two circumstances" (Arrighi, 1994, p. 30).

Those new rules would have *territory* and *borders* as key concepts also intended as new "fragile" certainties of the modern political system and of the rules of engagement of inter-national confrontation. This last aspect would be based on the mutual recognition of sovereign territories within certain boundaries.

Geography was no longer a system of organic and super-territorial vision of the world, as it had been in the Medieval past, but it embodied, in Modern Times, a new asset of political certainties, even if unstable, which substituted the medieval ones.

Contemporary uncertainties?

The current processes of globalization involving the political international structure, the boundaries and relations between States, would seem to represent, as proposed here and according to the thesis of Hardt and Negri (2000), further steps in a process which started with Modernity: the systemic uncertainty of the world after 1989 highlighted by Bolaffi and Marramao (2001), Colombo (2010; 2022), Hannay (2008), Jowitt (1992), Thrift[5] (1992) and Veca (1997) as well as by many other authors already cited would only be another step of *general crisis*[6] in world history, similar to globalization and the dynamics of Early Modern capitalism (Braudel, 1979; Hardt & Negri, 2000).

A comparison between the two epochs, even if many historians would see it as a hazardous one, can be put forward on the basis of the idea of a *general crisis* as a key concept both for Modernity[7] and for the current international political reality.[8] In both cases, a condition of deep *general crisis* seems to be produced, *de facto* coinciding with the idea of *uncertainty*.

Crisis seems to define the whole present epoch, characterized by elements of "chaos, terrorism, emergency, disorder and the obscure and instinctive Other; against order, reason, certainty, identity and nation" (Minca & Bialasiewicz, 2004, p. 226).[9] As stated by Hardt and Negri and as has been widely discussed earlier regarding the context of Modern Age, "Modernity itself is defined by crisis" (Hardt & Negri, 2000, p. 76). In both these periods, the typical phenomenon of general crisis, "the contraction of time and loss of boundaries" (Colombo, 2014, p. 12), took place.

The crisis that we experience today and that of Modernity are both *general crisis* characterized by instability of time (no one knows when it will end) and spatial extension (the geographical limits are also unknown). The present one is a *general crisis* because it goes beyond known boundaries, contributing to geographic indefiniteness and without clear definitions of the moment when a decision will be made to end the state of exception or emergency (Agamben, 2005; Schmitt, 2007). Only the starting moment – and not always – is clear in a *general crisis*, an awareness of its beginning exists. How it will evolve, how much space will be involved, how long it will last are all unknown. This spatial and temporal expansion confers the *status* of uncertainty to *general crises*, and they almost become synonymous.

General crises seem to characterize both Early Modernity, where medieval certainties of the old *modus vivendi* were questioned, and the current situation determined by the start of numerous *particular crises* that, as single critical events, can substantially be included within an overall crisis. In both cases, the result is an "indefinite, apparently unsolvable *condition*" (Colombo, 2014, p. 33). In the first scenario, because of the loss of an ordering principle such as the Imperial one, and in the second scenario, because of the fall of a system on which the Western and the Soviet world based their existences and cultures, as well as an international balance of power.

Both systems – the one derived from the bipolar contrast cited by Veca and the modern one born from capitalism studied by Schmitt and Aubert – followed certain, clear and established geopolitical configurations. The consequent uncertainty experienced by the two systems was the result of the fall of the previous assets based on religious (Gurevich, 1985; Le Goff, 2013) or ideological certainties. In particular, the post-Cold War world, coming "out of the disintegrating certainties of the Cold War" (Ó Tuathail & Luke, 1994, p. 382), was an "orphan of the certainties of bipolarity" (Minca & Bialasiewicz, 2004, p. 235).

If Early Modern Age territory had become the crucial and certain reference point able to substitute preceding certainties, in the globalized world that emerged from the Cold War, the territorial element was surpassed and a global, deterritorialized vision was affirmed which undermined the fragile certainties of the bipolar asset of the world.

The uncertainty of a post-bipolar system emerged in terms of *deterritorialization* (Ó Tuathail & Luke, 1994; Ó Tuathail et al., 1998) once a global dimension had been assumed, due to the emergence of supranational actors and of the world governance which progressively eroded the authority of Nation State authority. This is the other side of current geography of uncertainty: to this purpose, Veca makes reference to political, cultural and economic deterritorialization (1997).

The meeting point between the two systemic crises and the two geographical uncertainties is found in the territorial element that is so relevant in this analysis.

Veca perfectly explained what seems to have happened over time – from the 15th-century to the present day – to the territorial dimension in the current configuration of the geography of uncertainty: according to the author, currently

> the political, cultural and economic processes of deterritorialization, the clash of "civilizations", the erosion of resources of authority and sovereignty within the established boundaries, the gathering of international, transnational or regional institutions and organizations . . . modify the geopolitics of the established boundaries.
>
> (Veca, 1997, p. 226)

The uncertainty cited by Veca is therefore, at the same time, political and geographical and is related to the system following the global bipolar confrontation.

Carl Schmitt, focusing on the role of economics in the loss of territoriality of the Modern Age, stated that the economic processes had reduced the territory to an "empty space":

> territorial sovereignty was transformed into an empty space for socio-economic processes. The external territorial form with its linear boundaries was guaranteed, but not its substance, i.e., not the social and economic content of territorial integrity. The space of economic power determined the sphere of international law.
>
> (Schmitt, 2003, p. 252)

What should have been the focal point of the new certainties, European territoriality and a correspondence between national authority and specific territories, with certain borders defining public power, was already proving to have deep levels of uncertainty as an economic centrality in international relations. It is substantially the same idea expressed by Aubert about European modern geopolitics when speaking of "institutional fluidity of structures and diplomatic duties" as "only one aspect of a deeper and more general indeterminacy of institutional and bureaucratic boundaries of modern territorial States" (Aubert, 2008, p. 43).

The fact which seems to characterize both these historical moments, beyond the already cited end of past order which was able to guarantee a given system, even if unstable (both for Middle Age and the Cold War), is indeed the general crisis and what it generated.

The new chaotic structure arising from the Cold War opened the doors to a new phase of globalization in which there were "shifts that can be seen as moving towards deterritorialization and reterritorialization at the same time and in the same spaces" (Ó Tuathail & Luke, 1994, p. 382). The two authors have looked at the analysis of Gilles Deleuze and Félix Guattari (1987), who had made it clear, on the argument of *deterritorialization*, that "a set of artifices by which one element, itself deterritorialized, serves as a new territoriality for another, which has lost its territoriality as well" (Deleuze & Guattari, 1987, p. 174).

Besides the process of loss of territorialization, of the "shredding" of territoriality, established with the end of the two blocks system, there was a re-emergence of territorialization of single nationalities. This process which took place in Eastern Europe was known as re-territorialization (Khanna, 2011; Kaplan, 2012).

The fact that clearly emerges from the current political map of international relations[10] (see Luke & Toal, 1998), deeply undermined by uncertainty and redefinition of State and national identities, including ethnic-territorial identities, is that *deterritorialization* of recent years, at least where territorial affiliations have remained dormant and strong, has been followed by an attempt at *reterritorialization* and consequent reorganization of inter-State borders.

On consideration of some regions of the world, this would appear even more fitting and compelling and would confirm what has earlier been said. With the decline of a bipolar world, after a period of relative certainty compared to the necessary and inevitable unipolarity, that would have put an end to history (Fukuyama, 1992), the international system had to face, as Europe had done in the 16th-century, a period of instability generated by an attempt to supplant the order and certainty established within the two blocks by the claim and recognition of single Nation State individualities that had remained widely dormant and had been caused to remain "silent" by the cultural and ideological system of the two mutually opposed parts. It seems to be an attempt to regain an element of certainty, even if provisory, in such an extended reality from a geographic point of view, that it has lost its crucial points of reference.

The past 20 years have seen the end of United States' "innocence" after the Twin Towers attacks; the consequent ruinous wars in Afghanistan and Iraq, ruinous because uncertain in their objectives (it was a "global war on terror", therefore an undefinable object), and uncertain in their geographical extension ("global", therefore without specific borders); the affirmation of extra-State sovereignties and supranational organisms which superseded State decisions; the requests of global governance and the questioning of the role of Nation States, also because of financial giants; the recent economic crises that have caused American hegemony a further serious blow and an attempt to establish order after the Cold War; the "Arab Spring" and the affirmation of new actors, such as the Islamic State, on the international scenario; the pandemic crisis caused by COVID-19 and the consequent geopolitical and geoeconomic global redefinitions; and finally the recent war in Ukraine.

All this can be seen as geographic uncertainty emerging in all its disruptive force.

"All that is solid melts into air"

At the end of the 19th century, Marx noticed that "all that is solid melts into air". In the same way, Berman (1988) used this metaphor for the late 1980s to describe the modern way of life, and, after them, many have used the same ideas after the fall of the Berlin Wall. But these seem to be just the last and personal consequences of a common long-time dynamic, born during the Modern Age, when even Descartes noticed that everything is subject, as the wax, to a liquefaction process. It is what happened to human references (and this is the focal topic of sociologists), and to geographical and political divisions (which is the focal point of this book and of geographers).

All the distinctive features of current geography of uncertainty are evident in the vagueness of the inter-State boundaries, in the loss of certainties that had dominated past eras (in the Middle Ages as well as during the Cold War and in the period which immediately followed), in the affirmation of the principle of *deterritorialization* and in the resulting rise of nationalisms and territorial affiliations – in what has been defined as *reterritorialization*.

A systemic instability and insecurity (Colombo, 2022), which can be regarded as a *general crisis*, was also peculiar to the European geographic-political landscape in the 16th and 17th centuries, determined by the economic-capitalist dynamics and by the discovery of America that forced the European States to reassess, as highlighted by Hardt and Negri in their chapter on "Two Europes, Two Modernities" (Hardt & Negri, 2000).

***

At the conclusion of this itinerary, it is necessary to clarify whether uncertainty which started in Modernity has lasted to the present day, highlighting its most dramatically disruptive elements, allowing us to compare the idea of globalization to geography of uncertainty, and therefore to consider early modern globalization as the origin of present geographic uncertainty.

If in the Middle Ages, as we have tried to point out, knowledge of the world was based on a solid transcendental and religious vision which rested on ancient and universal assumptions, and if the advent of Modernity caused the liquefaction of that solidity – using the image provided by Marx and then Berman – and a paradoxical uncertainty which also affected the representative cartographic sphere, the further question that arises is: are we now experiencing the last offshoots of that world created by European openings to global spaces?

After their first and most impressive impetus in European Modernity and in its dynamics resulting from geographical discoveries, national and State singularities vehemently expressed themselves and made a virulent return in current inter-State relations, "out of the disintegrating certainties of the Cold War" (Ó Tuathail & Luke, 1994, p. 382).

In this dismantling process of established, certain structures, of deterritorialization and attempts to reterritorialize, in the uncertainty that characterizes relations among States, in the indeterminacy of global and inequal wars (Colombo, 2006), of erosion of the *jus ad bellum* (Benigno & Scuccimarra, 2007) as a symptom of the loss of centrality of Nation States, the condition of *general crisis* clearly appears and, with it, of *catastrophe* and *tragedy* of Modernity.

Indeed, "the catastrophe occurs when a power appears, that not only defines itself *autonomous* in principle from any actual spiritual value, but that does not even tolerate it" (Cacciari, 2013, p. 44): here, in this "catastrophe", in this basically *tragic* idea, it is possible to find the features of Modernity and of the present day, which expressed themselves in terms of geography of uncertainty, because exactly the same thing occurred in reality and in the perception of time. This also happened in the territorial field: "the universalistic vocation of imperial power, understood as an 'icon' of sovereignty, requires that time can be understood in the light of principles, contents and life forms that *stay*" (Ibid., p. 39), thus expressing a solidity, a stability, safety and certainty. The *imperial* power prior to Modernity – as the *ideological* power of the division of two sides in the 20th century – *stays*, and is contained within itself, is included in an all-encompassing and personalizing way, a conception of time and space in line with its principles, establishing itself *de facto* as an *Era*. In the *epoch*, the process of division and affirmation of multiplicities is required to experience, perceive and represent territorial space and reality according to its own logic, in accordance with its individual and relative value, such as to be configured within the concept of uncertainty.

In the *Era*, time and space are the expression of a universalistic tendency which conceives them as *eternal* and *infinite*, establishing the geographic and cartographic certainties of a system "that *stays*"; in the *epoch* – as intended here, linked with European Modernity and with the post-bipolar world system – the relativism expressed in the multiplicity of asserting individual *principles* and State entities causes the progressive loss of geographic certainties, in the multiplicity of expressions and in the possibility of claiming, on the basis of an affiliation and without accepting a superior power, more or less vast portions of the globe. This inevitably means that the *Era* is marked by the geographic (and of course temporal) certainty, while in the *epoch* of uncertainty, the map of the world only apparently establishes objective and truthful subdivisions hiding, in reality, turmoil and tumults in the territories, bringing them to the fore and towards the revenge of geography itself.

Indeed, "this continuous change [which is synonymous of uncertainty] manifests itself in the moving and continuously transforming network, differently intertwined, of property limits. This phenomenon can be conspicuously observed where these changes of limits occur in larger dimensions, in the coexistence of peoples, as we see it represented on geographic maps, which show us the historical dynamism in the relationships of power and possession of nations. But this continuous change is accompanied by a process that occurs rather internally ... From this comes the eternal rise and fall in the earthly organization of men" (Wust, 1985, pp. 116–117), which could coincide with geographical uncertainty.

From *Era* to *epoch*, from *stability* (what stays) to *instability*, from *certainty* to *uncertainty*: this seemed to be – in our discourse – the fate of human history of globalization, both in the individual and in the geographic sphere, that seems to recur from period to period, from historical moment to historical moment.

Because "all that is solid melts into air": all that was previously solidly established, all that once *stayed*, is sooner or later bound to dissolve in the air or, otherwise, to liquefy, like the wax of Descartes.

Notes

1 See Braudel, 1949; Rapp, 1975; Rosenzweig, 1917; Schmitt, 2015.
2 In the Christian religious vision, as opposed to the East, the universalist and centralist pretension was predominant: "even in the single situations in which the medieval Europe defined itself as West, for example distinguish itself from the byzantine East, remained untouched the idea of being the Christianity tout court; a universal pretention was advanced to comprehend the "world" in it" (Brunner, 1980, p. 37).
3 As seen in the previous chapter, this is the interpretation of many authors, who has observed in the "tragic" works of Shakespeare, Marlowe and other influential exponents of Modernity a new advent of the tragic element.
4 That is why we have distanced ourselves from the ideas of Khanna (2011) and Minc (1994), who wanted to see contemporary globalization as a current revival of the State multiplicity of the Middle Ages.
5 Who, by connecting the idea of an insecure world to the capitalistic dynamics trying to better understand them, especially looking at the work and managerial adjustments, said: "It is a discourse which depends on new metaphors, metaphors which attempt to capture a more turbulent, uncertain and insecure world" (Thrift, 2005, p. 32); and, again: "The new managerialism depends on the notion that the world is uncertain, complex, paradoxical, even chaotic (. . .). The manager must somehow find the means to steer a course in this fundamentally uncertain world, which she or he does by six main means" (Ibid., pp. 42–43).
6 Or a penultimate *systemic crisis*, if we consider the last years of world history, in which a new phase seems to have been started with the attack on the Twin Towers, the subsequent wars in Iraq and Afghanistan, the economic crisis and the latest geopolitical upheavals of 2014 and 2022 in Ukraine.
7 See Hobsbawm, 1954a; 1954b; Nexon, 2009; Parker & Smith, 1978; Trevor-Roper, 1967; Hardt & Negri, 2000.
8 See Colombo, 2014; Hardt & Negri, 2000.
9 Citing again the two authors, regarding the post-Cold War, "Huntington geography builds instead an ideal breeding ground for . . . the foreshadowing of a chaotic and dangerous world dominated by uncertainty and by completely unpredictable threats" (Minca & Bialasiewicz, 2004, p. 227).
10 See also, in this regard, Eastern Europe, and more specifically what is happening in Ukraine, with the redefinition of the State borders.

Bibliography

AA.VV. Comitato Nazionale per le Celebrazioni del V Centenario della Scoperta dell'America, *Cristoforo Colombo e l'apertura degli spazi*, Istituto Poligrafico Zecca dello Stato, Roma, 1992.

M. Aalbers, Geographies of the financial crisis, *Area*, 1, 2009, pp. 34–42.

K.A. Aastveit, G.J. Natvik and S. Sola, Economic uncertainty and the effectiveness of monetary policy, *Working Paper Norges Bank Research No. 17*, 2013.

G. Agamben, *State of Exception*, The University of Chicago Press, Chicago (IL)/London (UK), 2005.

G. Agamben, *The Omnibus Homo Sacer*, Stanford University Press, Stanford (CA), 2017.

J. Agnew, *Globalization and Sovereignty*, Rowman & Littlefield Publishers, Lanham (MD), 2009.

J. Agnew, *Making Political Geography*, Arnold, London (UK), 2002.

J. Agnew, The territorial trap: The geographical assumptions of international relations theory, *Review of International Political Economy*, 1(1), 1994, pp. 53–80.

A. Agumya and G.J. Hunter, Responding to the consequences of uncertainty in geographical data, *International Journal of Geographical Information Science*, 16(5), 2002, pp. 405–417.

M.W. Allan and V. Baláž, *Migration, Risk and Uncertainty*, Routledge, New York (NY), 2012.

S. Amin, *Eurocentrism: Modernity, Religion, and Democracy. A Critique of Eurocentrism and Culturalism*, Monthly Review Press, New York (NY), 2010.

S. Amin, G. Arrighi, A.G. Frank and I. Wallerstein, *Dynamics of Global Crisis*, Monthly Review Press, New York (NY), 1982.

M. Anam, S.H. Chiang and L. Hua, Uncertainty and international migration: An option cum portfolio model, *Journal of Labor Research*, 29(3), 2008, pp. 236–250.

J. Anderson, The shifting stage of politics: New medieval and postmodern territorialities? *Environment and Planning*, 6(14), 1996, pp. 133–154.

A. Appadurai, *Modernity at Large. Cultural Dimension of Globalization*, University of Minnesota Press, Minneapolis (MN)/London (UK), 1996.

R. Arnheim, *The Power of the Center. A Study of Composition in the Visual Arts*, University of California Press, Berkeley/Los Angeles (CA), 1982.

G. Arrighi, *The Long Twentieth Century. Money, Power and the Origins of Our Times*, Verso, London (UK), 1994.

G. Arrighi and B.J. Silver, Capitalism and world (dis)order, *Review of International Studies*, 27, 2001, pp. 257–279.

G. Arrighi and B.J. Silver, *Chaos and Governance in the Modern World System*, University of Minnesota Press, Minneapolis (MN)/London (UK), 1999.

T. Aston (ed.), *Crisis in Europe, 1560–1660. Essays from Past and Present*, Routledge & Kegan, London (UK), 1965.
P.M. Atkinson and G.M. Foody, *Uncertainty in Remote Sensing and GIS*, Wiley, Chichester, 2002.
A. Aubert, *L'Europa degli Imperi e degli Stati. Monarchie universali, equilibrio di potenza e pacifismi dal XV al XVII secolo*, Cacucci, Bari, 2008.
M. Augé, *Non-Places. Introduction to an Anthropology of Supermodernity*, Verso, London (UK)/New York (NY), 1995.
M. Azzari, *Natura e paesaggio nella Divina Commedia*, Phasar Edizioni, Firenze, 2012.
L. Bagrow, Old inventories of maps, *Imago Mundi*, 5, 1948, pp. 18–20.
B. Bailyn, *Atlantic History. Concept and Contours*, Harvard University Press, Cambridge (MA), 2005.
P. Barber, Medieval maps of the world, in P.D.A. Harvey (ed.), *The Hereford World Map. Medieval World Maps and their Context*, The British Library, London (UK), 2006.
P. Barber and T. Harper, *Magnificent Maps: Power, Propaganda and Art*, The British Library, London (UK), 2010.
V. Barbour, *Capitalism in Amsterdam in the 17th Century*, University of Michigan Press, Ann Arbor (MI), 1963.
J. Barnes, *Capitalism's World Disorder. Working-Class Politics at the Millennium*, Pathfinder, New York (NY)/London (UK)/Montreal/Sydney, 1999.
Z. Bauman, *Life in Fragments: Essays in Postmodern Morality*, Blackwell, Oxford/Cambridge (MA), 1995.
Z. Bauman, *Liquid Modernity*, Polity Press, Cambridge (MA), 2000.
U. Beck, *Risk Society. Towards a New Modernity*, Sage Publications, London (UK), 1992.
K. Beitel, Financial cycles and building booms: a supply side account, *Environment and Planning*, 12, 2000, pp. 2113–2132.
A. Béjin and E. Morin, Il concetto di crisi, in M. D'Eramo (ed.), *La crisi del concetto di crisi*, Lerici, Roma, 1980.
M. Bellotto and M. Foppolo (eds.), *L'incertezza. Workshop 1996. Indirizzo di Psicologia del Lavoro*, Ceup, Padova, 1997.
M. Benasayag and G. Schmit, *Les passions tristes. Suffrance psychique et crise sociale*, La Décuoverte, Paris, 2003.
F. Benigno and L. Scuccimarra (eds.), *Il governo dell'emergenza. Poteri straordinari e di guerra in Europa tra XVI e XX secolo*, Viella, Roma, 2007.
W. Benjamin, *Angelus Novus: Ausgewählte Schriften*, Suhrkamp, Frankfurt, 1966.
W. Benjamin, *The Arcades Project*, Harvard University Press, Cambridge (MA), 1999.
W. Benjamin, *The Origin of German Tragic Drama*, NLB, London (UK), 1977.
M. Berman, *All That Is Solid Melts Into Air. The experience of Modernity*, Penguin books, New York (NY), 1988.
S. Berninghaus and H.G. Seifort-Vogt, *International Migration Under Incomplete Information*, Springer Verlag, Berlin, 1991.
S. Bhagat and I. Obreja, Employment, corporate investment and cash flow uncertainty, *University of Colorado Working Paper*, 2013.
D. Birkholz, Hereford maps, Hereford lives: biography and cartography in an English cathedral city, in K.D. Lilley (ed.), *Mapping Medieval Geographies. Geographical Encounters in the Latin West and Beyond, 300–1600*, Cambridge University Press, Cambridge (MA), 2013.
M. Bloch, *Feudal Society*, University of Chicago Press, Chicago (IL), 1961.
H.P. Blossfeld, M. Mills and F. Bernandi (eds.), *Globalization, Uncertainty, and Men's Careers: An International Comparison*, Edward Elgar Publishing, Northampton (MA), 2006.

R. Bodei, Pensare il futuro, o dell'incertezza globale, *Lettera internazionale*, 106, 2010, pp. 1–4.
J. Bodin, *Les six livres de la Republique*, chez Lacques du Puys, Paris, 1579.
P. Boitani, *L'ombra di Ulisse*, il Mulino, Bologna, 1992.
A. Bolaffi and G. Marramao, *Frammento e sistema. Il conflitto-mondo da Sarajevo a Manhattan*, Donzelli, Roma 2001.
E. Boria, *Carte come armi. Geopolitica, cartografia, comunicazione*, Nuova Cultura, Roma, 2012.
E. Boria, *Cartografia e potere. Segni e rappresentazioni negli atlanti italiani del Novecento*, UTET, Torino, 2007.
G. Botero, *Le Relazioni Universali*, Nino Aragno Editore, Torino, 2015.
G. Botero, *The Reason of State*, Cambridge University Press, Cambridge (MA), 2017.
J. Branch, *The Cartographic State. Maps, Territory, and the Origins of Sovereignty*, Cambridge University Press, Cambridge (MA), 2014.
F. Braudel, *Afterthoughts on Material Civilization and Capitalism*, John Hopkins University Press, Baltimore (MD)/London (UK), 1979.
F. Braudel, *La Méditerranée et le monde méditerranéen à l'époque de Philippe II*, Armand Colin, Paris, 1949.
M. Brecher and J. Wilkenfeld, *Crisis, Conflict and Instability*, Pergamon Press, Oxford (UK), 1989.
J. Brotton, *A History of the World in 12 Maps*, Penguin, London (UK), 2012.
O. Brunner, *Sozialgeschichte Europas im Mittelalter*, Vandenhoeck & Ruprecht, Göttingen, 1978 [*Storia sociale dell'Europa nel medioevo*, il Mulino, Bologna, 1980].
D. Buisseret, *The Mapmakers' Quest. Depicting New Worlds in Renaissance Europe*, Oxford University Press, Oxford (UK), 2003.
H. Bull, *The Anarchical Society. A Study of Order in World Politics*, Palgrave, New York (NY), 2002.
J. Burckhardt, *The Civilization of the Renaissance in Italy*, The New American Library of World Literature, New York (NY), 1960.
M. Cacciari, *Il potere che frena. Saggio di teologia politica*, Adelphi, Milano, 2013.
F. Cantù (ed.), *Scoperta e conquista di un Mondo Nuovo*, Viella, Roma, 2007.
I. Caraci, *Al di là di altrove. Storia della geografia e delle esplorazioni*, Mursia, Milano, 2009.
I. Caraci, Nascita ed evoluzione della cartografia europea dell'America, in F. Cantù (ed.), *Scoperta e conquista di un Mondo Nuovo*, Viella, Roma, 2007.
P. Carta and R. Descendre (eds.), Géographie et politique au début de l'âge moderne, *Laboratoire Italien*, 8, 2008.
E. Casti, *L'ordine del mondo e la sua rappresentazione. Semiosi cartografica e autoreferenza*, Unicopli, Milano, 1998.
J. Casti, *Searching for Certainty: What Scientists Can Know About the Future*, Scribners, London (UK), 1991.
R. Cattedra and M. Memoli, Spazi di "nuova Italia": fra situazioni di cosmopolitismo urbano e condizioni di contenimento forzato, in A. Ricci (ed.), *Geografie dell'Italia molteplice. Univocità, economie e trasformazioni territoriali nel Mondo che cambia*, Società Geografica Italiana, Roma, 2013.
P.G. Cerny, Neomedievalism, civil war and the new. Security Dilemma: Globalisation as. Durable disorder, *Civil Wars*, 1(1), 1998, pp. 36–64.
P. Cesaretti (ed.), *È nella crisi che emerge il meglio di ognuno. Idee, numeri, racconti*, Bolis, Azzano San Paolo, 2013.

F. Chabod, *Scritti su Machiavelli*, Einaudi, Torino, 1964.
F. Chabod, *Storia dell'idea d'Europa*, Laterza, Roma-Bari, 2007.
S. Chan, *The End of Certainty: Towards a New Internationalism*, Zed Books, London (UK)/ New York (NY), 2009.
Y. Chun, M.P. Kwan and D.A. Griffith, Uncertainty and context in GIScience and geography: challenges in the era of geospatial big data, *International Journal of Geographical Information Science*, 2019, pp. 1131–1134.
G.L. Clark, Money flows like mercury: The geography of global finance, *Geografiska Annaler: Series B, Human Geography*, 87(2), 2005, pp. 99–112.
A. Colombo, *Il governo mondiale dell'emergenza. Dall'apoteosi della sicurezza all'epidemia dell'insicurezza*, Cortina Raffaello, Milano, 2022.
A. Colombo, *La disunità del mondo. Dopo il secolo globale*, Feltrinelli, Milano, 2010.
A. Colombo, *La guerra ineguale. Pace e violenza nel tramonto della società internazionale*, Il Mulino, Bologna, 2006.
A. Colombo, *Tempi decisivi. Natura e retorica delle crisi internazionali*, Feltrinelli, Milano, 2014.
H. Couclelis, The certainty of uncertainty: GIS and the limits of geographic knowledge, *Transactions in GIS*, 7(2), 2003, pp. 165–175.
J. Cowan, *A Mapmaker's Dream: The Meditations of Fra Mauro, Cartographer to the Court of Venice*, Shambhala, Boston (MA), 1996.
K.R. Cox, *Political Geography: Territory, State, and Society*, Blackwell, Oxford (UK), 2002.
K.R. Cox (ed.), *Spaces of Globalization*, Guilford Press, New York (NY), 1997.
J.W. Crampton and S. Elden (eds.), *Space, Knowledge and Power. Foucault and Geography*, Ashgate, Aldershot, 2007.
M. Czaika, Migration in times of uncertainty. On the role of economic prospects, *DEMIG Project Paper No. 11*, Oxford University Press, Oxford (UK), 2012.
M. D'Eramo (ed.), *La crisi del concetto di crisi*, Lerici, Roma, 1980.
S. Dalby, Critical geopolitics: Discourse, difference, and dissent, *Environment and Planning D: Society and Space*, 9(3), 1991, pp. 261–283.
P.G. Dalché, The Holy Land on medieval world maps, in P.D.A. Harvey (ed.), *The Hereford World Map. Medieval World Maps and their Context*, The British Library, London (UK), 2006.
E. Dardel, *L'Homme e la Terre*, Editions du CTHS, Paris, 1990.
G. Daverio Rocchi, *Frontiera e confini nella Grecia antica*, L'erma di Bretschneider, Roma, 1987.
M. De Cecco, *Gli anni dell'incertezza*, Laterza, Roma-Bari, 2007.
M. Dear, The Postmodern Challenge: Reconstructing Human Geography, *Transactions of the Institute of British Geographers* (New Series), 13(3), 1988, pp. 262–274.
G. Deleuze and F. Guattari, *A Thousand Plateaus: Capitalism and Schizophrenia*, University of Minnesota Press, Minneapolis (MN), 1987.
R. Descartes, *Discourse on Method and Meditations on First Philosophy*, Broadview Press, Peterborough, 2020.
R. Descendre, *Lo stato del mondo. Giovanni Botero tra ragion di Stato e geopolitica*, Viella, Roma, 2022.
J.H. Drèze, *Money and Uncertainty: Inflation, Interest, Indexation*, Edizioni dell'Elefante, Roma, 1993.
M. Ducza, Medieval world maps. Diagrams of a christian universe, *University of Melbourne Collections*, 12, 2013, pp. 8–13.
C. Dustmann, Return migration, uncertainty and precautionary savings, *Journal of Development Economics*, 52, 1997, pp. 295–316.

E. Edson, *Mapping Time and Space. How Medieval Mapmakers Viewed Their World*, The British Library, London (UK), 1997.

E. Edson, *The World Map, 1300–1492: The Persistence of Tradition and Transformation*, John Hopkins University Press, Baltimore (MD), 2007.

S. Elden, Reading Schmitt geopolitically. Nomos, territory and Großraum, *Radical Philosophy*, 161, 2010, pp. 18–26.

S. Elden, *The Birth of Territory*, The University of Chicago Press, Chicago (IL), 2013.

J.H. Elliott, *The Old World and the New. 1492–1650*, Cambridge University Press, Cambridge (MA), 1970.

P. Falchetta, *Fra' Mauro's World Map: A History*, Imago, Rimini, 2013.

P. Falchetta, Il "valore" dello spazio geografico: concezioni e percezioni tra le antiche mappaemundi e le rappresentazioni odierne, in M. Rossi (ed.), *Acta Concordia, Cartografia tra storia e web*, Fondazione Concordi, Rovigo, 2008.

M.A. Falchi Pellegrini, *Legittimità, legittimazione e resistenza nella teoria politica medioevale. Bartolo da Sassoferrato e Coluccio Salutati*, Ecig, Genova, 1981.

F. Farinelli, Genealogia del confine. Spazio geografico e spazio politico nella cultura europea, in AA.VV. Frontiere (ed.), *Politiche e mitologie dei confini europei*, Fondazione Collegio San Carlo, Modena, 2008.

F. Farinelli, *Geografia. Un'introduzione ai modelli del mondo*, Einaudi, Torino, 2003.

F. Farinelli, *I segni del mondo. Immagine cartografico e discorso geografico in età moderna*, La Nuova Italia, Firenze, 1992.

F. Farinelli, *La crisi della ragione cartografica*, Einaudi, Torino, 2009.

F. Fernández-Armesto, *1492. The Year the World Began*, Harper Collins, New York (NY), 2009.

G. Ferro, Cristoforo e Bartolomeo Colombo cartografi, in AA.VV. Cristoforo Colombo e l'apertura degli spazi (ed.), *Ufficio Centrale Per I Beni Librari E Gli Istituti Culturali*, Istituto Poligrafico e Zecca dello Stato, Libreria dello Stato, Roma, 1992.

G. Ferroni, *Machiavelli, o dell'incertezza*, Donzelli, Roma, 2003.

P. Fisher, Sorites paradox and vague geographies, *Fuzzy Sets and Systems*, 113, 2000, pp. 7–18.

G.M. Foody and P.M. Atkinson, *Uncertainty in Remote Sensing and GIS*, Wiley, Hoboken (NJ), 2002.

Foreign affairs, *The Age of Uncertainty*, 22(5), 2022.

M. Foucault, *Discipline and Punish. The Birth of the Prison*, Vintage books, New York (NY), 1995.

M. Foucault, *Security, Territory, Population: Lectures at the College De France, 1977–1978*, Palgrave MacMillan, London (UK), 2009.

T. Friedman, *The World Is Flat*, Farrar, Straus and Giroux, New York (NY), 2005.

F. Fukuyama, *The End of History and the Last Man*, Macmillan, New York (NY), 1992.

G. Fusco, M. Caglioni, K. Emsellem, M. Merad, Diego Moreno, et al. Questions of uncertainty in geography, *Environment and Planning A*, 49, 2017, pp. 2261–2280.

J.K. Galbraith, *The Age of Uncertainty*, Houghton Mifflin Company, Boston (MA), 1977.

C. Galli, Introduzione, in F. de Vitoria (ed.), *De Iure Belli* (C. Galli ed.), Laterza, Roma-Bari, 2005.

C. Galli, Presentazione, in C. Schmitt (ed.), *Amleto o Ecuba. L'irrompere del tempo nel gioco del dramma*, il Mulino, Bologna, 2012.

C. Galli, *Spazi politici. L'età moderna e l'età globale*, il Mulino, Bologna, 2001.

D. Gaumont, International migration and uncertainty: A non-factor price equalization overlapping generations model, *Brussels Economic Review*, 55, 2012, pp. 151–177.

C. Giacon, *La Seconda Scolastica*, Fratelli Bocca, Milano, 1950.
J. Gillies, *Shakespeare and the Geography of Difference*, Cambridge University Press, New York (NY), 1994.
R.J. Goebel, *A Companion to the Works of Walter Benjamin*, Camden House, Rochester (NY), 2009.
W. Goffart, *Historical Atlases. The First Three Hundred Years. 1570–1870*, The University of Chicago Press, Chicago (IL)/London (UK), 2003.
J.A. Goldstone, *Revolution and Rebellion in the Early Modern World*, University of California Press, Berkeley (CA), 1991.
I.R. Gordon and R.W. Vickerman, Opportunity, preference and constraint: An Approach to the analysis of metropolitan migration, *Urban Studies*, 19, 1982, pp. 247–261.
J. Gottmann, *The Significance of Territory*, The University Press of Virginia, Charlottesville (VA), 1973.
A. Gramsci, *Note sul Machiavelli, sulla politica e sullo Stato moderno*, Einaudi, Torino, 1949.
T. Gregory, Spazio sacro, spazio profano. I confini simbolici nel cristianesimo altomedievale, in AA.VV. Frontiere (ed.), *Politiche e mitologie dei confini europei*, Fondazione Collegio San Carlo, Modena, 2008.
A.J. Gurevich, *Categories of Medieval Culture*, Routledge, London (UK), 1985.
D. Hannay, *New World Disorder: The Un after the Cold War: An Insider's View*, I.B. Tauris, London (UK)/New York (NY), 2008.
M. Hardt and A. Negri, *Empire*, Harvard University Press, Cambridge (MA), 2000.
J.B. Harley, *The New Nature of Maps. Essays in the History of Cartography*, Johns Hopkins University Press, Baltimore (MD), 2001.
J.B. Harley and D. Woodward, *The History of Cartography* (Vol. I), The University of Chicago Press, Chicago (IL)/London (UK), 1987.
J.R. Harris and M.P. Todaro, Migration, unemployment, and development a two-sector analysis, *American Economic Review*, 60, 1970, pp. 126–142.
R.A. Hart, Inter-regional migration: Some theoretical considerations (Part 2), *Journal of Regional Science*, 15, 1975, pp. 289–305.
P.D.A. Harvey, *Mappa Mundi. The Hereford World Map*, Hereford Cathedral, Hereford, 2002.
P.D.A. Harvey, *Medieval Maps*, British Library, London (UK), 1991.
P.D.A. Harvey (ed.), *The Hereford World Map. Medieval World Maps and their Context*, The British Library, London (UK), 2006.
I. Hassan, *The Postmodern Turn*, Ohio State University Press, Columbus, 1987.
J.H. Helliott, *Imperial Spain, 1469–1716*, St. Martin's Press, New York (NY), 1964.
E.J. Hobsbawm, *The Age of Extremes. The Short Twentieth 1914–1991*, Abacus, London (UK), 1995.
E.J. Hobsbawm, The crisis of The 17th century – II, *Past and Present*, 6(1), 1954b, pp. 44–65.
E.J. Hobsbawm, The general crisis of the european economy in the 17th century, *Past and Present*, 5(1), 1954a, pp. 33–53.
B. Holsinger, *Neomedievalism, Neoconservatism, and the War on Terror*, Prickly Paradigm Press, Chicago (IL), 2007.
M. Horsman and A. Marshall, *After the Nation State: Citizens, Tribalism and the New World Disorder*, Harper Collins, London (UK), 1994.
C.T. Hunsaker, *Spatial Uncertainty in Ecology: Implications for Remote Sensing and GIS Applications*, Springer, New York (NY), 2001.
M. Isnardi Parente, Introduzione, in J. Bodin (ed.), *I sei libri dello Stato* (M. Isnardi Parente ed.), UTET, Torino, 1964.

ISPI, *L'età dell'incertezza. Scenari globali e l'Italia. Rapporto Annuale 2017*, Milano, 2017.

R.J. Johnston and P.J. Taylor, *A World in Crisis? Geographical Perspectives*, Blackwell, Oxford (UK), 1986.

K. Jowitt, *New World Disorder: The Leninist Extinction*, University of California Press, Berkeley (CA), 1992.

H. Kamen, *The Iron Century. Social Change in Europe, 1550–1660*, Praeger, New York (NY), 1971.

R. Kaplan, The coming anarchy: how scarcity, crime, overpopulation, tribalism, and disease are rapidly destroying the social fabric of our planet, *The Atlantic Monthly*, 273, 44 (21), 1994, pp. 1–16.

R. Kaplan, *The Revenge of Geography. What the Map Tells Us About Coming Conflicts and the Battle against Fate*, Random House, New York (NY), 2012.

P. Kennedy, *Preparing for the Twenty-First Century*, Random House, New York (NY), 1993.

P. Khanna, *How to Run the World. Charting a Course to the Next Renaissance*, Random House, New York (NY), 2011.

H. Kissinger, *Diplomacy*, Simon & Schuster, New York (NY), 1994.

H. Kissinger, *World Order*, Penguin Press, New York (NY), 2014.

M. Kitson, R. Martin and P. Tyler, The geographies of austerity, *Cambridge Journal of Regions, Economy and Society*, 4, 2011, pp. 289–302.

F.H. Knight, *Risk, Uncertainty, and Profit*, Hart, Schaffner & Marx, Houghton Mifflin Company, Boston (MA), 1921.

L. Kolakowsky, Uncertainties of a democratic age, *Journal of Democracy*, 1, 1990, pp. 47–50.

E.H. Kossmann, Discussion of H.R. Trevor-Roper: "The general crisis of the seventeenth century, *Past and Present*, 18, 1960, pp. 8–11.

D. Krasner, *Sovereignty: Organized Hypocrisy*, Princeton University Press, Princeton (NJ), 1999.

M. Kupfer, Reflection on the Ebstorf Map. Cartography, Theology and dilectio speculationis, in K. Lilley (ed.), *Mapping Medieval Geographies: Geographical Encounters in the Latin West and Beyond, 300–1600*, Cambridge University Press, Cambridge (MA), 2013.

L.R. Kurtz, *Evaluating Chicago Sociology: A Guide to the Literature*, The University of Chicago Press, Chicago (IL), 1984.

M.P. Kwan, The uncertain geographic context problem, *The Annals of the American Association of Geographers*, 102(5), 2012, pp. 958–968.

P.C. Langley, The spatial allocation of migrants in England and Wales: 1961–66, *Scottish Journal of Political Economy*, 21, 1974, pp. 259–277.

J. Le Goff, *L'imaginarie médiéval*, Gallimard, Paris, 2013 (ebook version).

E. Le Roy Ladurie, La crisi e lo storico, in M. D'Eramo (ed.), *La crisi del concetto di crisi*, Lerici, Roma, 1980.

R. Lensink, H. Bo and E. Sterken, Does uncertainty affect economic growth? An empirical analysis, *Weltwirtschaftliches Archiv*, 135(3), 1999, pp. 379–396.

G. Leti, *Teatro Belgico. O Vero Ritratti Historici, Chronologici, Politici e geografici delle sette province unite*, Apresso G. de Jonge, Amsterdamo, 1690.

K. Lilley (ed.), *Mapping Medieval Geographies: Geographical Encounters in the Latin West and Beyond, 300–1600*, Cambridge University Press, Cambridge (MA), 2013.

G. Linberg, *Urban Sociology School of Chicago: Researchers and Ideas 1892–1965*, Sveriges Sociologförbund, Lund, 2008.

D.V. Lindley, *Understanding Uncertainty*, Wiley, Hoboken (NJ), 2007.

C. Lois, *Terrae incognitae. Modos de pensar y mapear geografías desconocidas*, Eudeba, Buenos Aires, 2018.

A. Lombardo, *L'eroe tragico moderno. Faust, Amleto, Otello*, Donzelli, Roma, 2005.
A. Lombardo and E. Tarantino, *Storia del teatro inglese. L'età di Shakespeare*, Carocci, Roma, 2001.
N. Longo, *Studi Danteschi. Da Francesca alla Trincità*, Studium, Roma, 2013.
N. Luhmann, *Risk: A Sociological Theory*, de Gruyter, Berlin/New York (NY), 1993.
T. Luke and G. Toal, The fraying modern Map: Failed States and contraband capitalism, *Geopolitics*, 3(3), 1998, pp. 14–33.
G. Lunghini, *Conflitto Crisi Incertezza. La teoria economica dominante e le teorie alternative*, Bollati Boringhieri, Torino, 2012.
A. MacEachren, A. Robinson, S. Hopper, S. Gardner, R. Murray, M. Gahegan and E. Hetzler, Visualising geospatial information uncertainty: What we know and what we need to know, *Cartography and Geographic Information Science*, 32(3), 2005, pp. 139–160.
N. Machiavelli, *The Prince*, Oxford University Press, Oxford (UK), 2005.
N. Machiavelli, *The Prince and Discourses on Livy*, Restless Books, New York (NY), 2021.
M. Machina and K. Viscusi (eds.), *Handbook of the Economics of Risk and Uncertainty*, Elsevier/North Holland, Amsterdam, 2014.
G. Maier, Cumulation causation and selectivty in labour market oriented migration caused by imperfect information, *Regional Studies*, 19, 1985, pp. 231–241.
G. Mangani, *Il mondo di Abramo Ortelio. Misticismo geografia e collezionismo nel Rinascimento dei Paesi Bassi*, Franco Cosimo Panini, Modena, 2008.
G. Marramao, *The Passage West. Philosophy and Globalisation*, Verso, London (UK)/New York (NY), 2012.
C. Mazzei, *Il governo dell'incertezza. La pianificazione della città metropolitana tra globale e locale*, Grafill, Palermo, 2016.
F. Meinecke, *Machiavellism: The Doctrine of "raison d'État" and its Place in Modern History*, Routledge, New York (NY)/London (UK), 2017.
A. Minc, *Le nouveau Moyen Âge*, Gallimard, Paris, 1993 [*Il nuovo medioevo. Il declino della ragione in occidente*, Sperling&Kupfer, Milano, 1994].
C. Minca, *Spazi effimeri. Geografia e turismo tra moderno e postmoderno*, Cedam, Padova, 1996.
C. Minca and L. Bialasiewicz, *Spazio e Politica. Riflessioni di geografia critica*, Cedam, Padova, 2004.
A. Morris, Uncertainty in spatial databases, in John P. Wilson, A. Stewart Fotheringham (eds.), *The Handbook of Geographic Information Science*, Wiley, Hoboken (NJ), 2007.
S. Moruzzi, *Vaghezza. Confini, cumuli e paradossi*, Laterza, Roma-Bari, 2012.
H.T. Mowrer and R.G. Congalton, *Quantifying Spatial Uncertainty in Natural Resources: Theory and Applications for GIS and Remote Sensing*, Ann Arbor Press, Chelsea (MI), 2000.
H. Münkler, *Empires: The Logic of World Domination from Ancient Rome to the United States*, Polity, Cambridge (MA), 2007.
S. Nadler, *The Philosopher, The Priest, and the Painter. A Portrait of Descartes*. Princeton University Press, Princeton (NJ)/Oxford (UK), 2013.
D.H. Nexon, *The struggle for Power in Early Modern Europe: Religious Conflict, Dynastic Empires, and International Change*, Princeton University Press, Princeton (NJ), 2009.
C. Nicolet, *L'inventaire du Monde. Géographie et politique aux origines de l'Empire romain*, Fayard, Paris, 1988.
F.W. Nietzsche, *The birth of tragedy*, Oxford University Press, Oxford (UK), 2000.
R. O'Brien, *Global Financial Integration: The End of Geography*, Council on Foreign Relations Press, New York (NY), 1992.

P.G.J. O'Connell, Migration under uncertainty: "Try your luck" or "wait and see", *Journal of Regional Science*, 37(2), 1997, pp. 331–347.

G. Ó Tuathail, At the end of geopolitics? Reflections on a pluralizing problematic at the century's end, *Alternatives: Social Transformation and Humane Governance*, 22(1), 1997, pp. 35–55.

G. Ó Tuathail, Borderless worlds: Problematizing discourses of deterritorialization, *Global Finance and Digital Culture. Geopolitics*, 4(2), 1999.

G. Ó Tuathail, *Critical Geopolitics: The Politics of Writing Global Space*, University of Minnesota Press, Minneapolis (MN), 1996.

G. Ó Tuathail, S. Dalby and P. Routledge (eds.), *The Geopolitics Reader*, Routledge, London (UK), 1998.

G. Ó Tuathail and T. Luke, Present at (Dis)integration: Deterritorialization and reterritorialization in the new wor(l)d order, *Annals of the Association of American Geographers*, 84, 1994, pp. 381–398.

K. Ohmae, *The Borderless World: Power and Strategy in the Interlinked Economy*, HarperCollins, London (UK), 1990.

K. Ohmae, *The End of the Nation State: The Rise of Regional Economies*, HarperCollins, London (UK), 1995a.

K. Ohmae (ed.), *The Evolving Global Economy: Making Sense of the New World Order*, Harvard Business School Press, Boston (MA), 1995b.

A. Pagden, *Lords of All the World: Ideologies of Empire in Spain, Britain and France c.1500–c.1800*, Yale University Press, New Haven (CT), 1995.

M.P. Pagnini, *Geografia per il principe. Teoria e misura dello spazio geografico*, Unicopli, Milano, 1985.

G. Palumbo, Riannodare le fila della storia e della scienza, in F. Cantù (ed.), *Scoperta e conquista di un Mondo Nuovo*, Viella, Roma, 2007.

R.E. Park, E. Burgess and R. McKenzie, *The City*, University of Chicago Press, Chicago (IL), 1925.

G. Parker, *Europe in Crisis: 1598–1648*, Cornell University Press, Ithaca (NY), 1980.

G. Parker, *The World is Not Enough: The Imperial Vision of Philip II of Spain*, Markam Press Fund, Waco (TX), 2001.

G. Parker and L.M. Smith (eds.), *The General Crisis of the Seventeenth Century*, Routledge and Kegan Paul, London (UK)/Boston (MA), 1978.

T.V. Paul, J.J. Wirtz and M. Fortmann, *Balance of Power: Theory and Practice in the 21st Century*, Stanford University Press, Stanford (CA), 2004.

G. Pedullà, *Introduzione e Commento*, in N. Machiavelli, *Il Principe*, Donzelli, Roma, 2013.

L. Pirandello, *The Late Mattia Pascal*, New York Review Books, New York (NY), 2004.

M. Prak, *The Dutch Republic in the Seventeenth Century*, Cambridge University Press, Cambridge (MA), 2005.

M. Quaini, *Il mito di Atlante. Storia della cartografia occidentale in Età Moderna*, Il Portolano, Genova, 2006.

M. Quaini, L'età dell'evidenza cartografica. Una nuova visione del mondo fra Cinquecento e Seicento, in AA.VV. Comitato Nazionale per le Celebrazioni del V Centenario della Scoperta dell'America(ed.), *Cristoforo Colombo e l'apertura degli spazi*, Istituto Poligrafico Zecca dello Stato, Roma, 1992.

T.K. Rabb, *The Struggle for Stability in Early Modern Europe*, Oxford University Press, New York (NY), 1975.

C. Raffestin, La sfida della geografia tra poteri e mutamenti globali, *Documenti Geografici*, 2012, pp. 55–60.

C. Raffestin, *Pour une geographie du pouvoir*, Librairies techniques, Paris, 1980.

I. Ramonet, *Geopolitics of Chaos*, Algora, New York (NY), 1998.

F. Rampini, *L'età del caos*, Mondadori, Milano, 2015.

R.T. Rapp, The Unmaking of the mediterranean trade hegemony: International trade rivalry and the commercial revolution, *The Journal of Economic History*, 35(3), 1975, p. 499–525.

A.B. Raviola, *Giovanni Botero. Un profilo fra storia e storiografia*, Bruno Mondadori, Milano, 2020.

A.B. Raviola, Le Relazioni Universali di Giovanni Botero. Un viaggio politico nel mondo moderno, in G. Botero (ed.), *Le Relazioni Universali – Vol. 1*, Nino Aragno Editore, Torino, 2015.

N. Reed Kline, *Maps of Medieval thought. The Hereford Paradigm*, Boydell Press, Rochester (NY), 2001.

A. Ricci, Alle origini della geografia dell'incertezza. Il capitalismo mercantile nell'Olanda del Seicento, *Documenti Geografici*, 2010, 15, pp. 27–52.

A. Ricci, Challenges and revenge of borders. The Islamic state and Covid-19 as opposite poles of the same pendulum, in S. Zilli and G. Modaffari (eds.), *Confin(at)i/Bound(aries)*, Società di Studi Geografici. Memorie geografiche, Firenze, 2020a.

A. Ricci, Dalla crisi economico-finanziaria alla Geografia dell'incertezza. Mutamenti nel settore immobiliare e impatto sul territorio in alcune città italiane, *Documenti Geografici*, 1, 2013a, pp. 107–123.

A. Ricci, Globalizzazione, Riforma protestante e Secolarizzazione cartografica, Polemos, XI(2), 2018b, pp. 57–73.

A. Ricci, Il Principe *ovvero alle origini della geografia politica*, Società Geografica Italiana, Roma, 2015a.

A. Ricci, *La Geografia dell'Incertezza. Crisi di un modello e della sua rappresentazione in età moderna*, Exorma, Roma, 2017.

A. Ricci, La Geografia Globale dello Stato Islamico. Perché la mappa del Medio oriente (e non solo) sta cambiando, in AA.VV. Il Terrore che voleva farsi Stato (ed.), *Storie sull'Isis*, Eurilink, Roma, 2015b, pp. 79–11.

A. Ricci, La sfida delle migrazioni nella geografia dell'incertezza. Immagini e scenari geopolitici, *Rivista Geografica Italiana*, 4, 2020b, pp. 75–92.

A. Ricci, Lo Stato Islamico: sfida globale all'ordine geopolitico mondiale, *Rivista Trimestrale di Scienza dell'Amministrazione*, 3, 2018a, pp. 1–17.

A. Ricci, Machiavelli e la geografia dell'incertezza. Conoscenza del territorio e relazioni di potere nella modernità, in *Culture del testo e del documento le discipline del libro nelle biblioteche e negli archivi*, 2016, pp. 29–46.

A. Ricci, *Spazi di eccezione. Riflessioni geografiche su virus e libertà*, Castelvecchi, Roma, 2022.

A. Ricci, The affirmation of image and maps in the Modern Age: Cartographic secularization and protestant reformation, *Rendiconti Lincei. Scienze fisiche e naturali*, 32, 2021, pp. 45–55.

A. Ricci, The art of the art of geographical representation. Comparing Cartography and art in the Dutch Golden age, *Bollettino della Società Geografica Italiana*, XIII(VI), 2013b, pp. 655–677.

A. Ricci and C. Bilardi, *Cartografia, arte e potere tra Riforma e Controriforma. Il Palazzo Farnese a Caprarola*, Franco Cosimo Panini, Modena, 2020.

A. Ricci and F. Salvatori, Quale cartografia per una geografia dell'incertezza? *Bollettino della Associazione Italiana di Cartografia*, 165, 2019, pp. 122–132.

P.A. Rogerson, Spatial models of search, *Geographical Analysis*, 14, 1982, pp. 217–228.

M. Rosa (ed.), *Le origini dell'Europa moderna*, De Donato, Bari, 1977.

J.N. Rosenau, *Turbulence in World Politics: A Theory of Change and Continuity*, Princeton University Press, Princeton (NJ), 1990.
F. Rosenzweig, *Globus. Studien zur welgeschichtlichen Ranmlehre*, 1917.
M. Rossi, Comprendere il mondo dalla visione verticale a quella orizzontale, in M. Rossi (ed.), *Cartografie tra storia e web. Atti del convegno*, Fondazione Concordi, Rovigo, 2008.
G. Rossi, La governance globale e la fine della storia, *Il Sole 24 ore*, 27 luglio, 2014.
J.P. Roux, *Les explorateurs au Moyen Age*, Fayard, Paris, 1985.
J. Saarela and D.O. Rooth, Uncertainty and international return migration: some evidence from linked register data, *Applied Economics Letters*, 19, 2012, pp. 1893–1897.
F. Salvatori, *Il mondo nuovo, I nuovi mondi. Paolo Emilio Taviani e gli studi su Colombo*, Società Geografica Italiana, Roma, 2013.
F. Sanguineti, *Gramsci e Machiavelli*, Laterza, Roma-Bari, 1981.
S. Sassen, *Territory, Authority, Rights: From Medieval to Global Assemblage*, Princeton University Press, Princeton (NJ), 2006.
A. Scafi, *Alla scoperta del paradiso. Un atlante del cielo sulla terra*, Sellerio, Palermo, 2011.
A. Scafi, Defining mappaemundi, in P.D.A. Harvey (ed.), *The Hereford World Map. Medieval World Maps and their Context*, The British Library, London (UK), 2006a.
A. Scafi, *Mapping paradise: A History of Heaven on Earth*, University of Chicago Press, Chicago (IL) 2006b.
A. Scafi, *Maps of Paradise*, British Library, London (UK), 2013.
S. Schama, *The Embarrassment of Riches. An Interpretation of Dutch Culture in the Golden Age*, University of California Press, Berkeley (CA), 1988.
A. Schiavone, Limes. La politica dei confini dell'Impero romano, in AA.VV. Frontiere (ed.), *Politiche e mitologie dei confini europei*, Fondazione Collegio San Carlo, Modena, 2008.
C. Schmitt, *Hamlet or Hecuba. The Irruption of Time into Play*, Plutarch Press, Corvallis (OR), 2006.
C. Schmitt, *Land and Sea. A World-Historical Meditation*, Telos Press Publishing, Candor (NY), 2015.
C. Schmitt, *Political Theology, Four Chapters on the Concept of Sovereignty*, University of Chicago Press, Chicago (IL), 2005.
C. Schmitt, *The Concept of the Political*, University of Chicago Press, Chicago (IL), 2007.
C. Schmitt, *The Nomos of the Earth in the International Law of the Jus Publicum Europaeum*, Telos Press Publishing, New York (NY), 2003.
R.A. Schwartz, J.A. Byrne and A. Colaninno (eds.), *Volatility: Risk and Uncertainty in Financial Markets*, Springer, New York (NY), 2011.
I. Scoones, What is uncertainty and why does it matter? *STEPS Working Paper No. 105*, ESRC STEPS Centre, 2019.
N. Senanayake, B. King, Geographies of uncertainty, *Geoforum*, 123, 2021, pp. 129–135.
A. Sestini, *Cartografia Generale*, Pàtron, Bologna, 1981.
M. Shapiro, Moral geographies and the ethics of post-sovereignty, *Public Culture*, 6(3), 1994, pp. 479–502.
M. Shapiro, Sovereignty and exchange in the orders of Modernity, *Alternatives: Global, Local, Political*, 16(4), 1991, pp. 447–477.
A. Shattuck, Toxic uncertainties and epistemic emergence: Understanding pesticides and health in Lao PDR, *The Annals of the American Association of Geographers*, 2020, pp. 1–15.

E. Sheppard and H. Leitner, Quo vadis neoliberalism? The remaking of global capitalist governance after the Washington Consensus, *Geoforum*, 41, 2010, pp. 185–194.

W.Z. Shi, Modeling uncertainty in geographic information and analysis, *Science China Technological Sciences*, 51, 2008, pp. 38–47.

R. Silvey, Development geography: Politics and 'the state' under crisis, *Progress in Human Geography*, 6, 2010, pp. 828–834.

R. Simek, *Heaven and Earth in the Middle Ages: The Physical World Before Columbus*, Boydell & Brewer, Suffolk/Rochester (NY), 1996.

G. Simmel, *The Philosophy of Money*, Routledge, London (UK)/New York (NY), 2004.

P. Sloterdijk, *In the World Interior of Capital. For a Philosophical Theory of Globalization*, Polity Press, Cambridge (MA), 2013.

N. Smith, Global economic crisis and the need for an international critical geography, *Korean Journal for Space and Environment*, 12, 1999, pp. 37–65.

N. Smith, *The Endgame of Globalization*, Routledge, London (UK), 2005.

F. Somaini, *Geografie politiche italiane tra Medio Evo e Rinascimento*, Officina Libraria, Milano, 2013.

M. Storper, Territories, flows, and hierarchies in the global economy, in K. Cox (ed.), *Spaces of Globalization: Reasserting the Power of the Local*, Guilford Press, New York (NY), 1997, pp. 19–44.

J.C. Thill and D.Z. Sui, mental maps and fuzziness in space preferences, *The Professional Geographer*, 1993, pp. 264–276.

N. Thrift, *Knowing Capitalism*, Sage Publications, London (UK), 2005.

N. Thrift, *New World Disorder*, University of California Press, Berkeley (CA), 1992.

N. Thrift, The geography of international economic disorder, in R.J. Johnston and P.J. Taylor (ed.), *A World in Crisis?* Blackwell, Oxford (UK), 1989.

N. Thrift, The rise of soft capitalism, *Cultural Values*, 1(1), 2007, pp. 29–57.

C. Tilly, *European Revolutions, 1492–1992*, Blackwell, Cambridge (MA), 1993.

C. Tilly, States and nationalism in Europe 1492–1992, *Theory and Society*, 23(1), 1994, pp. 131–146.

M.P. Todaro, A model of labor migration and urban unemployment in less developed Countries, *American Economic Review*, 59, 1969, pp. 138–148.

F. Tönnies, *Gemeinschaft und Gesellschaft: Grundbegriffe der reinen Soziologie*, Karl Curtius, Berlin, 1922.

H. Trevor-Roper, *The Crisis of the Seventeenth Century. Religion, the Reformation and Social Change*, Liberty Fund, Indianapolis (IN), 1967.

H. Trevor-Roper, The general crisis of the 17th century, *Past and Present*, 16, 1959, pp. 31–64.

A. Turco, *Configurazioni della territorialità*, Franco Angeli, Milano, 2010.

A. Turco, *Paesaggio: Pratiche linguaggi mondi*, Diabasis, Reggio Emilia, 2002.

L. Urbani Ulivi, Saggio introduttivo, in R. Descartes (ed.), *Meditazioni metafisiche*, Bompiani, Milano, 2001.

V. Valerio, *La Geografia di Tolomeo e la nascita della moderna rappresentazione dello spazio*, in AA.VV (ed.), *Cartografia e istituzioni in età moderna*, Società Ligure di Storia Patria, Genova, 2012.

A. Vallega, *Ecumene Oceano. Il mare nella civiltà: ieri, oggi, domani*, Milano, Mursia, 1985.

G.E. Valori, *Geopolitica dell'incertezza*, Rubbettino, Roma, 2017.

142 Bibliography

M.P.R. van den Broecke, *Ortelius Atlas Maps. An illustrated Guide*, Hes Publishers, Westrenen, 1996.

C. Van Duzer, *Sea monsters on Medieval and Renaissance Maps*, The British Library, London (UK), 2013.

S. Veca, *Dell'incertezza. Tre meditazioni filosofiche*, Feltrinelli, Milano, 1997.

A. Vespucci, *Letters From a New World: Amerigo Vespucci's Discovery of America* (ed. L. Formisano), Marsilio, New York (NY), 1992.

A. Vespucci, *Mundus Novus. Letter to Lorenzo Pietro di Medici*, Princeton University Press, Princeton (PJ), 1916.

A.D. von den Brincken, Jerusalem on medieval mappaemundi: a site both historical and eschatological, in P.D.A. Harvey (ed.), *The Hereford World Map*, The British Library, London (UK), 2006.

I. Wallerstein, *After Liberalism*, The New Press, New York (NY), 1995.

I. Wallerstein, Crisis as transition, in S. Amin, G. Arrighi, A.G. Frank and I. Wallerstein (eds.), *Dynamics of Global Crisis*, Monthly Review Press, New York (NY), 1982.

I. Wallerstein, *The Modern World-System I-II-III*, University of California Press, Berkeley (CA)/Los Angeles (CA)/London (UK), 2011.

T. Wang and T.S. Wirjanto, The role of risk and risk aversion in an individual's migration decision, *Stochastic Models*, 20, 2004, pp. 129–147.

T. Wang and T.S. Wirjanto, The role of risk aversion and uncertainty in individual's migration decision, *Working Paper University of Waterloo No. 98003*, 1997.

W.E. Washburn, Il significato della "Scoperta" nei secoli XV e XVI, in F. Cantù (ed.), *Scoperta e conquista di un Mondo Nuovo*, Viella, Roma 2007.

I. Watt, *Myths of Modern Individualism: Faust, don Quixote, don Juan, Robinson Crusoe*, Cambridge University Press, Cambridge (MA), 1996.

M. Weber, *Economy and Society. An Outline of Interpretive Sociology*, University of California Press, Berkley/Los Angeles (CA), 1978.

A. Wellmer, On the dialectic of modernism and postmodernism, *Praxis International*, 4, 1985, pp. 337–362.

S.D. Westrem, *The Hereford World Map: A Transcription and Translation of the Legends with Commentary*, Folio Society, London (UK), 2010.

M. Wight, *System of States*, Leicester University Press, Leicester, 1977.

T. Windholz, *Strategies for Handling Spatial Uncertainty due to Discretization*, PhD Thesis, The University of Maine, Maine, 2001.

L. Wirth, *The Ghetto*, University of Chicago Press, Chicago (IL), 1928.

P. Withfield, *The Image of the World. 20 Centuries of World Maps*, British Library, London (UK), 1994.

D. Wood, *The Power of Maps*, Guilford Press, New York (NY), 1992.

D. Woodward, Medieval Mappaemundi, in J.B. Harley and D. Woodward (eds.), *The History of Cartography* (Vol. I), The University of Chicago Press, Chicago (IL)/London (UK), 1987b.

J.K. Wright, The geographical lore of the time of the crusades: A study in the history of medieval science and tradition in Western Europe, *American Geographical Society Research*, 15, 1925, pp. 489–543.

P. Wust, *Naivität und Pietät*, J.C.B. Mohr, Tübingen, 1925.

P. Wust, *Ungewissheit und Wagnis*, Standort, München, 1940 [*Rischio e Incertezza*, Morcelliana, Brescia, 1985].

H.W.C. Yeung, Capital, state and space: contesting the borderless world, *Transactions of the Institute of British Geographers*, 23, 1998, pp. 291–309.

P. Zanini, *Significati del confine. I limiti naturali, storici, mentali*, Bruno Mondadori, Milano, 2000.
J. Zhang and M.F. Goodchild, *Uncertainty in Geographical Information*, Taylor and Francis, London (UK), 2002.
J.O. Zinn (ed.), *Social Theories of Risk and Uncertainty: An Introduction*, Blackwell, Malden (MA), 2008.
P. Zurla, *Il Mappamondo di Fra Mauro Camaldolese*, Venezia, 1806.

Index

1492 81
1517 111
1555 13, 46
1648 8, 13, 18n, 79, 122
1989 123
2007–2008 viii, 26, 31, 40n
9/11 viii

Africa IX, 15, 62, 80
Agamben, G. xvii, 45, 65, 124
age of discovery xiv, 3, 14, 17, 76, 77–8, 100; great discoveries 64, 69, 77, 122; great geographical discoveries 9, 10, 19n, 42, 61, 85, 94, 102, 109, 110, 114, 115, 119n; geographical discoveries 10, 15, 52, 79, 114, 127
Age of Uncertainty, the (Galbraith's work) xii
Agnew, J. xiii, xviii, 2, 26–7, 39n, 65, 67n
America viii, ix, 12, 14, 15, 21, 25, 53, 76–8, 81, 82n, 103, 106, 119, 126–7; discovery of 17, 19n, 52, 54, 84, 99, 100, 119n, 121
Amin, S. xviiin, 16, 17, 19n
anarchy 51; anarchical society 48, 123
Appadurai, A. xiii, 25
Arab Spring ix, 126
Aristotelian 64, 72
Aristotle 53, 57
Arnheim, R. 91
Arrighi, G. xii, xiii, xviiin, 8, 9, 13, 17, 18n, 19n, 25, 28–9, 36, 39n, 61, 123
Asia 15, 62, 76–7, 80, 101n, 105
Atlantic 15, 19n, 63, 77, 78, 81n, 101n
atlas (collection of maps) 49, 102, 103, 106, 120n; *Atlas* (mythological figure) 103, 104
Aubert, A. 41 42, 43, 44, 55, 61, 66, 124, 125
Augé, M. xiii, 39n

Augsburg *see* Peace of Augsburg
Azzari, M. 71

balance of power x, 10, 41, 45, 47, 48, 49, 124
Barbour, V. 78
Bauman, Z. xiii, xviiin, 19n, 23, 24, 39n, 65
Beck, U. xiii, 23, 39n, 65
Beitel, K. 32
Benjamin, W. 25, 31, 39n, 120n
Berlin Wall 31; fall of the 7, 17, 23, 26, 127
Berman, M. xviiin, 19n, 24, 65, 127
bipolarity 28, 124; bipolar system ix, 8, 9, 17, 18, 25, 26, 28, 29, 31, 124, 125, 126; post-bipolar 125, 128
Bodei, R. 29, 30, 31
Bodin, J. 44, 53, 54, 55, 56, 57, 67n
Boitani, Piero 69, 71, 72, 73, 75
Bolaffi, A. 25, 123n
border vi, 10, 20, 61, 62, 68n, 86, xiii, xv, xvi, 6, 7, 10, 11, 12, 13, 14, 16, 18, 36, 40n, 41, 43, 48, 50, 54, 55, 60, 61, 62, 63, 64, 65, 68n, 70, 72, 78, 81, 86, 96, 102, 118, 119, 123, 125, 126, 129n
borderless world xiii, 61, 67n
Boria, E. xvii, 118
Botero, G. x, 44, 50, 51, 53, 56, 67n
boundary ix, 4, 6, 7, 41, 42, 48, 50, 51, 52, 55, 57, 60, 61, 62, 63, 64, 65, 66, 68n, 70, 71, 72, 84, 86, 97, 122, 123, 124, 125, 127
Branch, J. 1, 79, 85, 100n
Braudel, F. 18n, 36, 77, 78, 82n, 123, 129n
Brunner, O. 19n, 28, 66n, 123, 129n
Burckhardt, J. 11

Cacciari, M. 128
Caliphate ix; Islamic State ix, 126

capitalism xii, 20, 65, 124; early modern 78, 123
Caraci, I. 15, 119n
Carta, P. 41
cartography i, xv, xviiin; 1, 3, 16, 70, 76, 79, 80, 83, 84, 85, 86, 89, 94, 99, 100, 100n, 102, 105, 106, 108, 110, 114, 116, 117, 118, 119n, 120n, 122; medieval 3, 80, 86, 91, 92, 97, 100, 102, 117; modern 15, 25, 84, 88, 94, 101n 106; cartographic reason 83; cartographic representation *see* representation; cartographic secularization 3, 98, 122
Casti, E. x, 2, 70, 71, 74, 84, 85
Cattedra, R. xiv, xviii, 34, 35, 40n
centre; loss of 84, 107; loss of centrality 16, 109, 110, 112, 128
certainty vi, viii, x, xii, xviiin, 4, 5, 6, 10, 14, 16–17, 23, 26–8, 30, 31, 37, 39, 46, 48, 54–5, 58, 59, 60, 62, 63, 65, 70, 72, 77, 79, 80–1, 83, 86–7, 89, 90–2, 94, 98, 103–5, 109, 110–11, 113–14, 116, 119, 124, 126, 128; geographical 72; of representation 16, 27, 39, 70, 83, 116
Cervantes, de M. 119
Chabod, F. 13, 19n, 44, 46, 47, 49, 62, 66n
chaos ix, xiii, xv, 1, 7, 8, 9, 13, 17, 23–4, 26–9, 34–6, 41, 49, 51, 89, 107, 121, 124; systemic 9, 13, 123
Charlemagne 54, 67n
Charles V 50, 53–4, 63
Chicago 22; School of *see* School of Chicago
Christ 90–4, 97, 99, 103, 112, 117
Church 11, 19n, 28, 57, 66n, 80, 98, 100n, 101n, 117
City 22, 35, 38, 95, 111, 112, 119; cities xiv, 11, 32–5, 40n, 43, 47, 49, 89; modern cities 23–4; holy city (Jerusalem) 91
civil war 13
Clark, G. xiii, 31–2
Clausewitz, von C. 49
cogito ergo sum 57, 109, 116; *je pens pense, donc je suis* 60
Cold War ix, xii, xiii, 8, 17–18, 23, 26–9, 31, 39n, 124–7; post–Cold War viii, ix, 6, 9, 18, 27, 29, 36, 124, 129n

Colombo, A. xviin, xviiin, 6, 7, 9, 18n, 19n, 25–6, 54, 61, 65–6, 68n, 119n, 123–4, 127, 129n
Columbus C. x, 14, 17, 46, 58, 70, 71–6, 80–1, 85, 100, 119
Counter-Reformation 6
Cox, K.R. xvi, 44, 65
crisis, of the 17th century 19; economic ix, 7, 26, 31–2, 81, 129n; general crisis xv, 7, 9, 10, 11, 13, 17–18, 87, 123–8; global xi, 115; of cartographic reason 83
Cuius Regio, Eius Religio 46

Dante Alighieri xv, 37, 69, 71–3, 75, 80
Dardel, E. 70–1
de Lapradelle, P. 63
De Principatibus see The Prince
de' Medici, L. x
de' Medici, Lorenzo di Pier Francesco 74
Descartes, R. 27, 44, 54, 56–9, 60, 62, 64–7n, 86, 127, 129
Descendre, R. 41, 50
deterritorialization *see* territorialization
diplomacy 39n, 42, 47; *Diplomacy* (Kissinger's work) 48; diplomats 49
Discourses on Livy (Machiavelli's work) 46
discovery of America *see* America
discoveries; geographical discoveries; great geographical discoveries *see* Age of Discovery
disorder xi, xii, xiii, 2, 9, 25–6, 28–9, 31, 36, 39n, 49, 121, 124
disorientation 107, 140
Don Quixote 115, 119
Dustmann, C. 33, 40n

Early Modern Age i, 70, 124; Early Modern Europe 43, 100n; Early Modern globalization i, x, 127; Early Modern Times 29; Early Modernity i, xv, 16–17, 19n, 28, 124
Ebstorf *mappamundi* 90, *90*, 91–3, 99, 101n
economic crisis *see* crisis
Eden 80
Elden, S. 5, 12, 28, 51
emergency viii, xviiin, 65, 124
empire 50, 54–5, 62–3, 66n, 67n, 77; *Empire* (Hardt & Negri's work) 10; roman 52, 63, 68n; imperial logic 55–6, 65, 68n; imperial vision 5; imperial power 67n, 128; imperial

borders 55, 65; imperialism xii, 12, 27
England 106–7, 109
Eurasia 13, 26
Eurocentrism 16–17; European centrality 14, 16–17, 97, 122
Europe xv, 8, 10–16, 19n, 26–7, 43–4, 46–9, 51–2, 54, 56–7, 62–4, 66, 69, 74, 76–9, 80–1, 87–8, 100, 103, 105, 107, 110–12, 118, 121, 125–7, 129n; Modern Europe 42, 45, 112, 123; Medieval Europe 74, 129n; Old Continent 12–14, 27, 41, 51, 63, 69, 78–9, 110, 121–2
exception, state of 65, 124
explorers vi, 37, 69, 70, 74–5, 80–1, 113

Falchetta, P. 94, 96
Farinelli, F. vii, xvii, 2, 11, 15–16, 76, 83–4, 90, 97–8, 101n, 103
Faust(us), Dr. 81n, 83–4, 111–14, 119, 120n
Fernández-Armesto, F. 70
Ferro, G. 119n
Ferroni, G. 45–6
feudal, asset 9; system 5–6, 18, 106; world 52
Foucault, M. 5–7, 18
Fra Mauro *mappamundi* 94, *95*, 96–8, 101n, 120n
France 56
frontiers 15, 29, 63
Fukuyama, F. 9, 29, 31, 39n, 126

Galbraith, J.K. xii, xiii
Galli, C. 17, 53, 107–9, 121
geographer in full sail *see* geography in full sail
geographic information system (GIS) xviiin, 18n, 20–1
geographical determinism 55
geographical representation 18n, 20–1, 98
geography in full sail 70, 74–5; geographer in full sail 84, 111, 113–14, 119
geography of uncertainty i, xiv, xv, 3, 4, 8, 16, 18, 25, 34–5, 39, 44, 51, 56, 61, 71, 74, 76, 80–1, 121–2, 125, 127–8
geopolitics i, xiii, 4, 25, 125; geopolitical situation i, xiii
Giacon, C. 52, 67n
GIS *see* geographic information system
global linear thinking 12, 50, 74
global lines 122

globality 16–17, 74, 76, 79, 81, 113, 115, 122
globalization i, viii, ix, x, xi, xv, xvi, 10, 17–18, 22, 25–6, 28, 30, 32, 36, 39, 61, 65, 74, 81, 123, 126–7; contemporary 129n; economic 32, 68n; modern i, x, 127; political 68n
globus 70
God 24, 37, 50, 59, 60, 67n, 71, 75, 80, 87–8, 91–3, 98, 101n, 102–3, 117
Gottmann, J. 61, 63, 65
Gramsci, A. 50, 66
Gurevich, A.J. 55, 66, 80, 86–9, 90, 92, 101n, 123–4

Hamlet (character) 50, 81n, 83–4, 106–14, 117–18; *Hamlet* (Shakespear's work) 110, 11, 114
Hardt, M. xviiin, 10–13, 16–17, 123–4, 127, 129
Harley, J.B. 1, 82n, 86–7, 115, 120n
Harvey, P.D.A. 11, 88, 90, 101n
heaven 11, 89, 92, 96–7, 113
Hellenism 62
Hereford *mappamundi* 68n, 92, *93*
hero, tragic 71–2, 81, 108
Hobbes, T. 47
Hobsbawm, E. xii, xviiin, 25, 28, 39n, 129n
Holsinger, B. 28–9
horror vacui 80, 82n, 92, 97
human geography vii

Iago 114–16, 118
immanence 12–13, 55, 57–8, 63, 83, 102, 104–5, 108
imperialism *see* empire
indecision 3–7, 9, 18, 108, 11, 114
indeterminacy 7, 10, 42, 60, 125, 127
indeterminateness 46, 48
individualism 59, 111–12, 120n
insecuritas 36–8
insecurity 3–7, 9, 18, 23, 36, 52, 127
international politics xii, xvi, 6, 7, 36, 52
international relations i, ix, x, xiii, xiv, xv, 4, 14–15, 25–9, 32, 43–4, 49, 50, 61, 103, 121, 125–6
inter-State, borders 18, 126; boundaries 122, 127; relations xv, 41–2, 127
irrevocable reality 50–1, 105, 109–10
Islamic State *see* Caliphate
Isnardi Parente, M. 54–6, 67n
Italy i, 26, 35, 40n, 45, 47–9

Jerusalem 78, 80, 88, 91–2, 95, 97–8, 102, 116; *Jerusalem delivered* 72
Jowitt, K. xiii, xviiin, 19n, 39n, 61, 123
Jus Publicum Europaeum 47, 65, 122–3
justum bellum 53

Kaplan, R. xviiin, 10, 11, 54
Khanna, P. xviiin, 19n, 25, 28–9, 39n, 126, 129n
Kissinger, H. xiii, 8, 41, 48–9
Knight, F. xii, 31

laboratory geography 70, 84; laboratory geographers 114
landscape 46, 100, 101n, 115; political 46, 123, 127; diplomatic 39n
Le Goff, J. 86, 90–1, 123–4
limes 63, 68n
Liquidity 42, 61
Lois, C. 76, 85
Lombardo, A. 36, 72, 81n, 83, 89, 104–10, 112, 114–17, 120n
loss of centre (or centrality) *see* centre
Louis XIV 42
Lucretius 46
Luke, T. 124–7
Luther, M. 111

Machiavelli, N. x, 6, 44–9, 50–1, 53, 57, 66–7
Mad Flight (Ulysses') 15, 37, 58, 69, 71–2, 74–5, 80–1, 85, 96, 113
Magellan, F. x
Mappamundi, mappaemundi 3, 16, 80, 88, 89, 90, 92, 94, 97, 100, 101n, 102, 105, 116, 118; of Fra Mauro *see* Fra Mauro; of Hereford *see* Hereford; of Ebstorf *see* Ebstorf
maps, medieval 62, 80, 86–9, 91, 99; modern 85–6; maps and power i
Marco Polo, *The Travels of* 96
Marlowe, C. 81n, 111–12, 114, 120n, 129n
Marramao, G. 25, 74, 123
Marx, K. 24, 65, 127
Mediterranean 15–16, 19n, 63, 77–8, 81, 81n, 82n, 122
Memoli, M. xiv, xviii, 34, 35, 40n
Mercator, G. 102–4, 111, 117, 120n; projection 110, 118
Middle Age (or Middle Ages) xiv, 8, 11, 14, 16, 19n, 27–9, 39n, 42, 48, 54, 56, 58, 62, 66, 66n, 67n, 71–2, 78–9, 86–9, 94–6, 99, 100, 100n, 101n, 106–7, 109, 112, 116, 118, 120n, 123, 125, 127, 129n; new Middle Age *see* Neomedievalism; medieval cartography *see* cartography; medieval maps *see* maps; medieval mentality 63, 90, 94; medieval period 80, 85; medieval vision 96, 98, 110; medieval way of life 66n
migrations 33, 36; migrant flows 35; migratory flows xiv, 33
Minc, A. 26–9, 31, 39n, 129n
Minca, C. xiii, xiv, xviii, 28, 124, 129
Modernity i, x, xiv, xv, 3, 5, 8, 9, 10–17, 19n, 24, 28, 30, 39, 41, 44, 47, 51, 54, 57, 59, 60–2, 64, 72–3, 81, 81n, 83–5, 87, 94–5, 100, 101n, 102–6, 108–12, 114–19, 121–4, 127–8, 129n; Modern Age i, v, x, xiv, xv, xviiin, 3, 8, 9, 11, 14, 18, 19n, 25, 28, 38, 41, 44, 46, 54–5, 57, 61–3, 65–6, 69–72, 74, 76, 79, 81, 84, 86–7, 105–7, 112, 114–15, 117, 120n, 124–5, 127; modern cartography *see* cartography; modern era 22, 57, 65, 69, 79, 80; modern geography 100, 121; modern maps *see* maps; modern mentality 86, 97; modern period ix, 61, 74, 80, 85–6, 121; Modern Times xv, xviiin, 8, 29, 70, 94, 101n, 103, 110, 113, 117, 123; modern vision 105; modern way of life 127
money 31–2
Moruzzi, S. 6, 36
multiplicity 10, 28, 41, 44, 48, 54–5, 68n, 128, 129n
multipolarity 44; multi-polarity 25
Mundus Novus 63, 69

Nadler, S. 57–8
Nation State x, xii, 6, 25, 29, 48, 61, 64, 68n, 85, 125–6, 128; National State 61
nationalism xiii, 26, 127
nationality 123
Negri, A. xviiin, 10–13, 16–17, 123–4, 127, 129
neomedievalism 28; new Middle Ages 27–9, 39n
Netherlands, the i, viii, 22, 111; Dutch Golden Age i, viii, 102; United Provinces 54, 102

Index

New World 10–12, 17, 19n, 51, 53, 63, 69, 74, 76–9, 86–7, 100n, 103, 115, 119, 120n
new world order x, xi, 115
Nicolet, C. 2
Nietzsche, F.W. 24, 83
nomos of the Earth 107–8
non plus ultra 63, 68n
Non-places xiii, 39n

Ó Tuathail, G. xiii, xviiin, 2, 39n, 61, 124–7
O'Connell, P.G.J. 33
ocean 74, 77; oceanic routes 15; oceanic space 74, 103, 119n
Odysseus *see* Ulysses
Ohmae, K. xiii, xviiin, 61
Old Continent *see* Europe
order, global viii, ix, 6, 8–9, 17, 26; new world order *see* new world order; world viii, 27, 91, 106, 108
organic view, organic State
Ortelius, A. 102–4, 106, 117, 120n
Othello 81n, 83–4, 105, 114–16, 118

paradise 89, 96–7, 103, 107
Parker, G. xviii, 11, 50, 54, 65–6, 129n
Peace (or Treaty) of Westphalia 13, 79
Peace of Augsburg 13, 46
Pedullà, G. 45–8, 66n
Pillars of Hercules 12, 63, 68n, 72, 75, 77, 96
Pindar 72
Pirandello, L. 83, 94
place xviiin, 3–4, 15, 22, 33, 35, 49, 61, 63, 67n, 87–9, 91–2, 96, 98–9, 103, 113, 117–18
Plato 62
political geography i, ix, xiv, xvi, 25, 28, 41–2, 44, 50–1, 57, 67n, 87, 109, 111
Post-modernity 64
power, balance of *see* balance of power
precariousness xiii, 116
Protestant Reformation 14, 98, 111; Reformation 6, 11, 12, 14, 52, 54–5, 109, 112, 119, 120n
Ptolemy 1, 99, 100, 101n, 104, 111

Quaini, M. 102–4, 118, 120n

Raffestin, C. 2, 50–1, 85–6
raison d'État 48, 50

Ramonet, I. xiii, 26
realism 3, 45–6, 51, 88
Reed Kline, N. 82n, 101n
relativism 36, 109, 128; geographical 115
Renaissance x, 11, 12, 14, 16, 19n, 41–2, 66n, 78, 84, 97, 106, 117, 120n
representation, cartographic x, xv, 1, 16, 79, 80–1, 83, 85, 88, 94, 97, 99, 102, 104, 110, 119; geographic 102
reterritorialization *see* territorialization
revenge of geography xiii, 26, 128
revolution xv, 3, 7–11, 13–15 19n, 41, 44, 51, 54, 58, 66, 69, 70, 74, 78, 83, 96–8, 110–11, 114, 117, 119n, 122; geographical 14, 70; spatial *see* space
Rey, J.-M. 30
risk vii, 23, 28, 34, 36, 38, 119
Roman Age *see* Rome
Rome i, viii, xi, xvii, xviiin, 22, 62, 78, 103; Roman Age 62
Rooth, D.-O. 34
Rosenzweig, F. 129n
Rossi, G. 29
Rossi, M. 89
Roux, J.P. 14, 17

Saarela, J. 34
safety 7, 14, 17, 23, 46, 49, 128; unsafety 23
Salamanca 53
Salvatori, F. viii, xvii, xviiin, 75
Sanguineti, F. 50
Scafi, A. 86, 92, 102n
Schama, S. 54, 102
Schmitt, C. 4, 12, 13, 15, 35, 47, 50, 65, 70, 72, 74, 78, 83–4, 102–9, 121–2, 124–5, 129n
School of Chicago 22, 23, 39n, 65
Second Scholasticism 44, 51, 52, 53, 67n; Scholastic thinkers 58
secularization 12, 98–100, 117; cartographic *see* cartography
security 5, 6, 13, 47, 73, 113; insecurity *see* insecurity
Shakespeare, W. 49, 81n, 88, 104, 106–7, 109, 112, 116, 119, 129
Shapiro, M. xiii, xviiin
Silver, B.J. xii, xiii, xiiin, 8, 13, 17, 18n, 19n, 25, 39n, 61
Simmel, G. 31, 55
Sloterdijk, P. xviiin, 12, 63

Smith M. xviiin, 54, 65, 66, 129n
Smith, N. 26, 32
solidity 14, 127–8
sovereignty 6, 41–2, 46–7, 53–5, 85, 125, 128
space xiii, xiv, 2, 3, 5, 7, 12, 15, 16, 21, 23, 27, 29, 35, 37, 39n, 40n, 49–51, 61, 65, 67n, 69, 70, 72, 76, 78, 79, 81, 82n, 85, 86, 89, 91, 92, 94, 98, 103–4, 107–8, 110, 113, 119, 119n, 122, 124–6, 128; geographical 2, 32; global i, xiv, 11, 16, 74, 127; spatial revolution 54, 121; spatial order 79, 122
stasis (στάσις) 45
State 5, 41–4, 47, 50–2, 54, 55, 61, 63, 65, 68n, 85, 86, 100n, 107, 123, 126–8, 129n
Subjectivity 12, 52, 64, 113–16
System of States 51

Tasso, T. 72
terrae incognitae 76
territorial trap xiii, 26, 27
territorialization 126; deterritorialization xiii, 6, 125–7; reterritorialization 126–7
territory i, vi, 5, 6, 18, 22, 32–3, 43, 46–7, 50–1, 53, 61, 62, 66n, 67n, 68n, 77, 85, 123–5
terrorism 124
The Prince (Machiavelli's work) 6, 46–7, 49–51, 67n; *De Principatibus* 46
theatrum mundi 49, 88, 100, 102–6, 116, 119
Thrift, N. xiii, xviiin, 19n, 20, 30, 39n, 61, 123, 129n
Tocqueville, de A. 30
Tönnies, F. 22
tragedy, modern x, 94, 110–11, 119; cartographic 25, 83–4
transcendence 12, 55, 97, 105, 108; transcendental vision 55, 75, 98, 113, 116
travellers 69, 70, 84, 99, 113–14, 117

Trevor-Roper, H. xviiin, 11, 19n, 54, 65–6, 129n
tribalism 26, 31
Turco, A. xv, 14, 15, 19n, 66n

Ukraine, 7, 126, 129n; Russo-Ukrainian war xi
Ulysses 37, 69, 71–5, 113; Odysseus 71
Uncertainty, cartographic 103; geographical xv, 17, 41, 69, 86, 128; global x, 6, 18, 28–9; modern x, xi, xv, 12, 14, 25, 27, 31, 37–8, 44, 56–7, 60, 68n, 73, 79, 87, 105–6, 108, 117–18
unipolarity 29, 126
United Provinces *see* Netherlands, the
United Provinces, Netherlands
United States of America viii, 9, 26, 29, 36, 126
universalism 12, 14, 26, 55–6, 58, 123
unsafety *see* safety
Urbani Ulivi, L. 59, 60, 64, 67n

vagueness 2, 4, 6, 7, 9, 11, 18, 21–2, 27, 36, 68n, 127
Valerio, V. 99
Vallega, A. 15
Veca, S. xiii, 6, 8, 17, 26, 36, 123–5
Vespucci, A. x, 69, 74, 75, 78, 100
Vitoria de, F. 52–3

Wallerstein, I. 8, 9, 25, 36, 81n
war on terror ix, 26, 28, 126
wax (metaphor of) 62, 64–6, 86–7, 127, 129
Weber, M. 22, 30
West 19n, 78, 122, 129n
Westphalia, peace of *see* Peace of Westphalia
Wittenberg 111–13, 119
Woodward, D. 82n, 86, 89, 101n
world system 68n, 102
Wust, P. 5, 36–8, 57, 63, 73, 128

Yeung, H.W.C. xiii, 20, 61, 67n

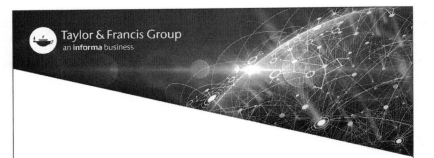